G000047682

SMALL-SCALE TEXTILES

FABRIC MANUFACTURE

Other books in this series:

Spinning
Dyeing and Printing
Yarn Preparation

SMALL-SCALE TEXTILES

FABRIC MANUFACTURE

A handbook

Alan Newton

Intermediate Technology Publications 1993

Intermediate Technology Publications
103/105 Southampton Row
London WC1B 4HH, UK

© Intermediate Technology Publications 1993

A CIP catalogue record for this book is available from the British Library

ISBN 1-85339-133-6

Printed by Russell Press Ltd, Nottingham, UK

CONTENTS

		Page
Acknowledgements		vi
Foreword		vii
Preface		viii
Chapter 1	Introduction to fabric manufacture	1
Chapter 2	Basic principles and processes of fabric manufacture	7
Chapter 3	Simple methods of determining the quality of fabrics	26
Chapter 4	Types of hand-looms and knitting machines	32
Chapter 5	Planning for production	37
Chapter 6	Equipment suppliers	42
Chapter 7	Sources of further information	43

Appendices

1 Woven cotton fabric specification

2 Comparison of Tex with other count systems

3 Cloth (worsted) setting chart : yarn count/cm

4 Cloth (cotton) setting chart : yarn count/in

5 Glossary

ACKNOWLEDGEMENTS

Writing a book of this kind demands contributions from many people. To acknowledge all those whose knowledge and time have been put to use would be almost impossible here. However, thanks must go to Michael Henderson who drafted the first manuscript, and Tom Kilbride who revised it. Some of this text is included more or less verbatim. Thanks also to John Foulds and Martin Hardingham who masterminded the design of the handbook, and to Ethan Danielson, whose illustrations are informative and artistic.

Alan Newton

FOREWORD

This handbook is one of a series dealing with small-scale textile production, from raw materials to finished products. Each handbook sets out to give some of the options available to existing or potential producers, where their aims could be to create employment or sustain existing textile production, the ultimate goal being to generate income for the rural poor in developing countries. Needless to say, this slim volume does not pretend to be comprehensive. It is intended as an introduction to the topic which may stimulate further enquiry. Although each handbook is complete in itself and provides useful reference material on each specific area of production, the series, taken as a whole, reveals the breadth of technology required to equip a small-scale textile industry. While being primarily technical, the series also covers some of the socio-economic, managerial, and marketing issues relevant to textile production in the rural areas of developing countries.

The production of this series of books has been sponsored by Intermediate Technology (IT) UK, as part of its efforts to help co-ordinate the most appropriate solutions to particular development needs. The series forms part of the cycle of identifying the need, recognizing the problems, and developing strategies to alleviate the crisis of un- and under-employment in the developing world.

IT also offers consultancy and technical enquiry services. For further information write to IT, Myson House, Railway Terrace, Rugby, Warks. CV21 3HT, UK. We will be pleased to help.

Martin Hardingham
IT, Rugby, UK.

PREFACE

Textile fabrics have been constructed in different ways for thousands of years. Early fabrics were produced and used as protection against the elements, chiefly as garments or coverings of one sort or another. Now textiles are also used for decorative purposes, as domestic furnishings for example, and for many industrial uses, where a strong, flexible material is required. It was soon realized that assembling fibres into yarns and fabrics was an effective way to produce textiles which possessed the properties of warmth, the ability to drape, be strong and to last for a long time while in constant use.

Numerous methods of making different fabric structures have been devised across the world, and since the nineteenth century, complex machinery has been developed to make these structures cheaply and at high speed. At the same time, many people are making fabrics by hand in small-scale cottage industries, such as woollen hand-knitted garments in cold climates, cotton and silk handloom weaving in India and the making of fishermen's nets in the islands of the South Pacific. Textile fabric production may be carried out as a creative hobby, for home use, to supply fabrics for use at work, or for sale in the textile market as a means of obtaining income. Although there is rarely a choice regarding the actual method of fabric manufacture; say, between weaving or knitting, there can be a choice regarding the 'appropriate' level of machinery to be used in a particular situation. A particular type of fabric can often be made with either a simple or a more complex machine, but one is rarely a direct alternative to the other. Complex machines are usually more difficult to manufacture, maintain, or repair, cost more, take up more space and have a more limited function. On the other hand, they usually produce textiles faster, increase the options for added value, are physically easier to use, requiring less skill and less human involvement. Although the large-scale production of textiles in highly-automated factories has made available a large volume of cheap fabrics, thus benefiting people all over the world, rural communities in developing countries can benefit directly by producing their own fabrics with the help of simple hand processes or small-scale technologies, provided the economic balance between raw material availability, production costs and suitable markets can be achieved. Quite often the benefits are obtained by creating new employment opportunities, perhaps in addition to other activities, such as small-scale farming. Fabric making is then undertaken to supplement existing incomes and increase very low earnings.

This handbook is concerned with fabric production as a part of small-scale textile production as a whole, and is aimed at small, rural communities and the development organizations or individual field-workers involved with them. Quite often the need is for some basic information or understanding, either to start a new textile operation, or to improve an existing one.

Alan Newton

1. INTRODUCTION TO FABRIC MANUFACTURE

This handbook is an introduction to simple methods of making types of textile fabrics used throughout the world. It is intended for readers interested in small-scale processes and therefore some techniques are only briefly mentioned. There are six basic structures that are used to form fabrics. It is convenient to arrange the various structures and methods of making fabric in decreasing order of importance and usage:

- ☐ weaving (interlacing)
- ☐ knitting (looping)
- ☐ fibre entanglement (felting)
- ☐ lace-making (twining)
- ☐ braiding
- ☐ knotting (netting)

Any one of these may be highly important locally. All on this list, except felting, use yarns which have been produced in some previous yarn-production process. Felting creates fabric by the entanglement of loose fibres. This produces materials which are generally rather different from the conventional textile fabrics, stiff and lacking the ability to drape, or low in strength, such as felt hat material or carpet underlay. Normally however, fabric production is carried out with yarn as the basic raw material.

FABRIC CHARACTERISTICS

The characteristics of a textile fabric depend chiefly on three considerations.

Yarn. Including the type of fibre, the arrangement of fibre in the yarn, the colour, thickness and amount of twist.

Method of arranging yarns. By weaving, knitting, twining, braiding and knotting.

Density of yarns in the fabric. The combination of the number of yarns and the thickness of the yarns governs the weight of the fabric and its cover. For instance, fabrics for furnishings or sacks are likely to be heavier than fabrics for coats or blankets, which are heavier than fabrics for shirts or sheets.

While fabrics for mosquito nets are more open (*low cover*) than fabrics for tarpaulins (*high cover*), most fabrics will have an intermediate value of cover, generally quite high but not so tightly constructed that they have a stiff handle.

FABRIC STRUCTURES

The examples that follow show fabric structures which are characteristic of the type of fabric being described. The path of each yarn through the structure should be observed carefully in order to understand how the fabric is formed.

Woven fabrics

Woven fabrics are constructed with two sets of interlacing yarns, the warp yarns, or ends, which lie along the length of the fabric, and the weft yarns, or picks, which are inserted one after the other and lie across the fabric, interlacing with the warp. The variety of arrangements of yarns in a woven structure is very large, but a few are used more often. Illustration 1 shows four different weaves.

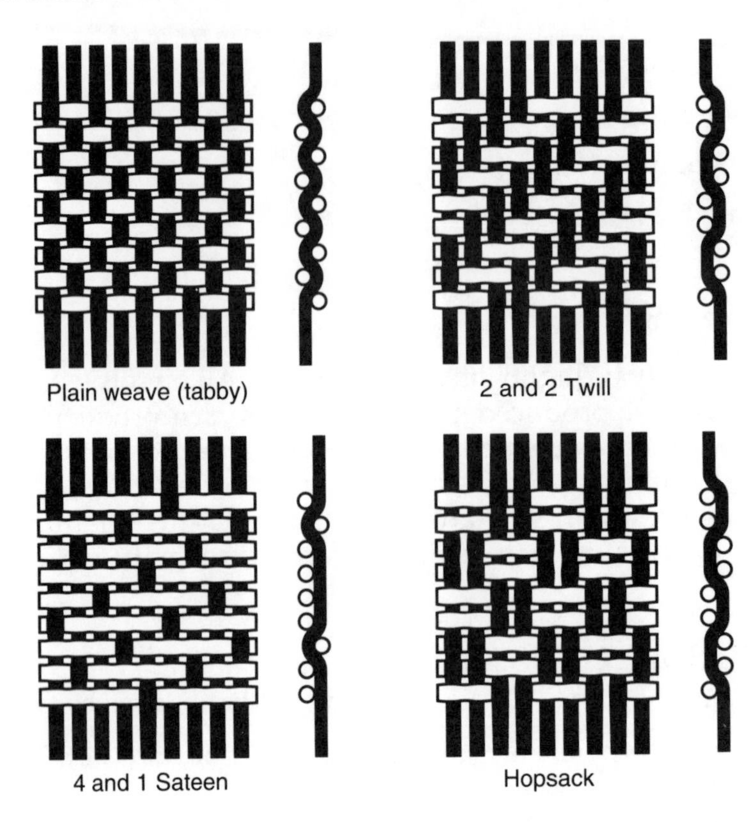

*Illustration 1 **Basic weaves***

Plain, or tabby, weave is the most common and the simplest method of interlacing. In plain weave the warp yarns pass over and under every weft yarn in succession. Twill, hopsack, satin (warp float) and sateen (weft float) weaves contain 'floats' of yarn, where the one yarn may pass over several yarns before interlacing. Although plain weave, with its high number of interlacings is the firmest weave, it is not necessarily the most suitable for all purposes. The main reason for changing the structure is to achieve the best combination of weight and cover for the eventual use of the fabric.

Knitted fabrics

Knitted fabrics are made from interlocking loops of yarn. There are two distinct types of knitted fabric, weft-knit, so-called because the yarns lie across the fabric, as in the structures shown in Illustration 2, and warp-knit, in which the yarn path is basically down the length of the fabric, as can be seen in the structure shown in Illustration 3.

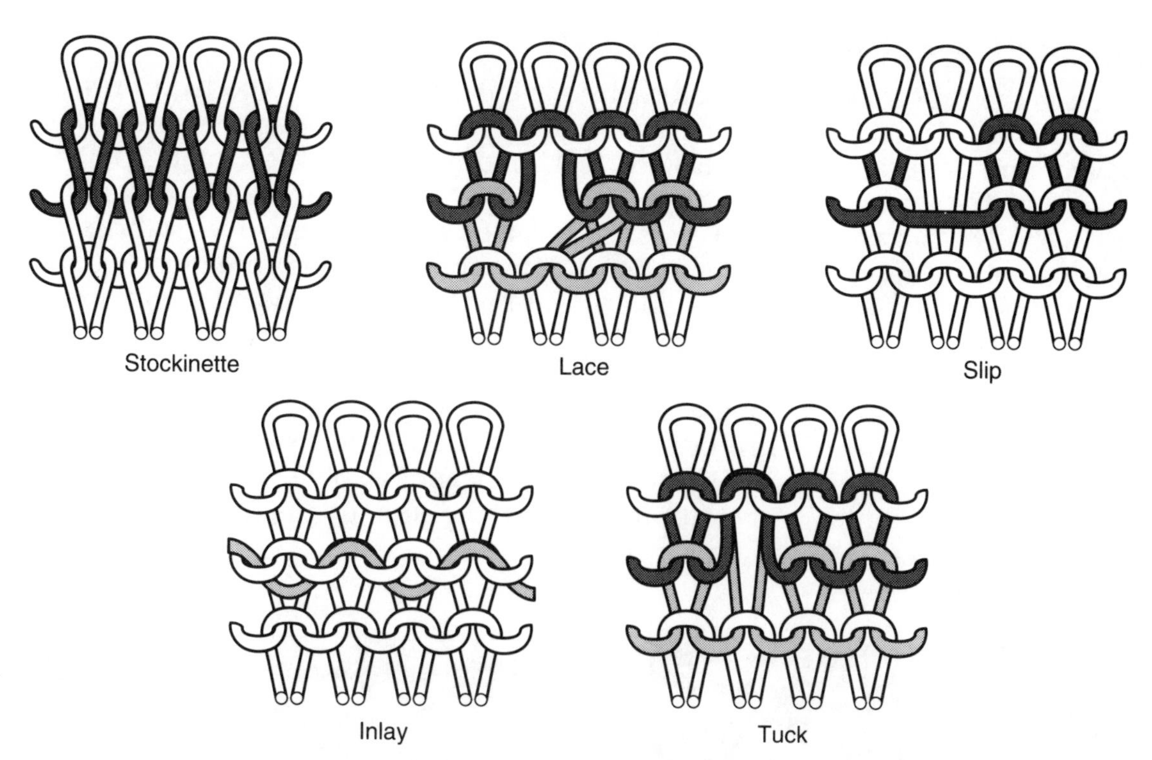

Illustration 2 Basic weft-knit stitches

Illustration 3 Warp knit - locknit stitch

Each row of stitches is called a 'course', and each column is called a 'wale'. In machine knitting, each single wale of stitches will be made on a separate needle. Knitted fabrics made with knitting needles, by hand or on domestic knitting machines, are usually weft-knit. Weft-knit structures are easily deformed, so they stretch and relax easily, and they are bulky, so this makes them ideal for garments. Weft-knitting differs from weaving and warp-knitting in that the weft-knit fabrics can be made from a single yarn package. Warp-knit fabrics are almost invariably made on complex machines for a limited number of end uses. Warp-knit fabrics have complicated structures in which the yarns are less capable of movement than those in weft-knit fabrics and, therefore, have less stretch.

Felts and nonwovens

Felting is the process of making fabric by the entanglement of woollen fibres. All animal fibres, of which wool is one, have a scaly surface, and when a mass of them is given a mechanical action under hot, wet conditions, the fibres entangle. Other fabrics which do not contain a proportion of animal fibres can be produced from fibre webs by 'nonwoven' processes. The chief types of nonwoven fabrics are adhesive-bonded fabrics (see illustration 4a), in which the fibres are held together by a 'binder', such as synthetic rubber, heat-bonded fabrics using a mixture of man-made fibres with different melting points (see Illustration 4b), and needle-punched fabrics, in which the fibres have been entangled by barbed needles (see Illustration 4c). All these nonwoven fabrics have special uses such as interlinings and stiffeners for garments, disposable diapers or nappies, tea bags, bandages, hats, filters and carpets. Their chief characteristic is cheapness, obtained by high-speed output on high capital cost equipment, and the aim in making them is the achievement of special physical properties.

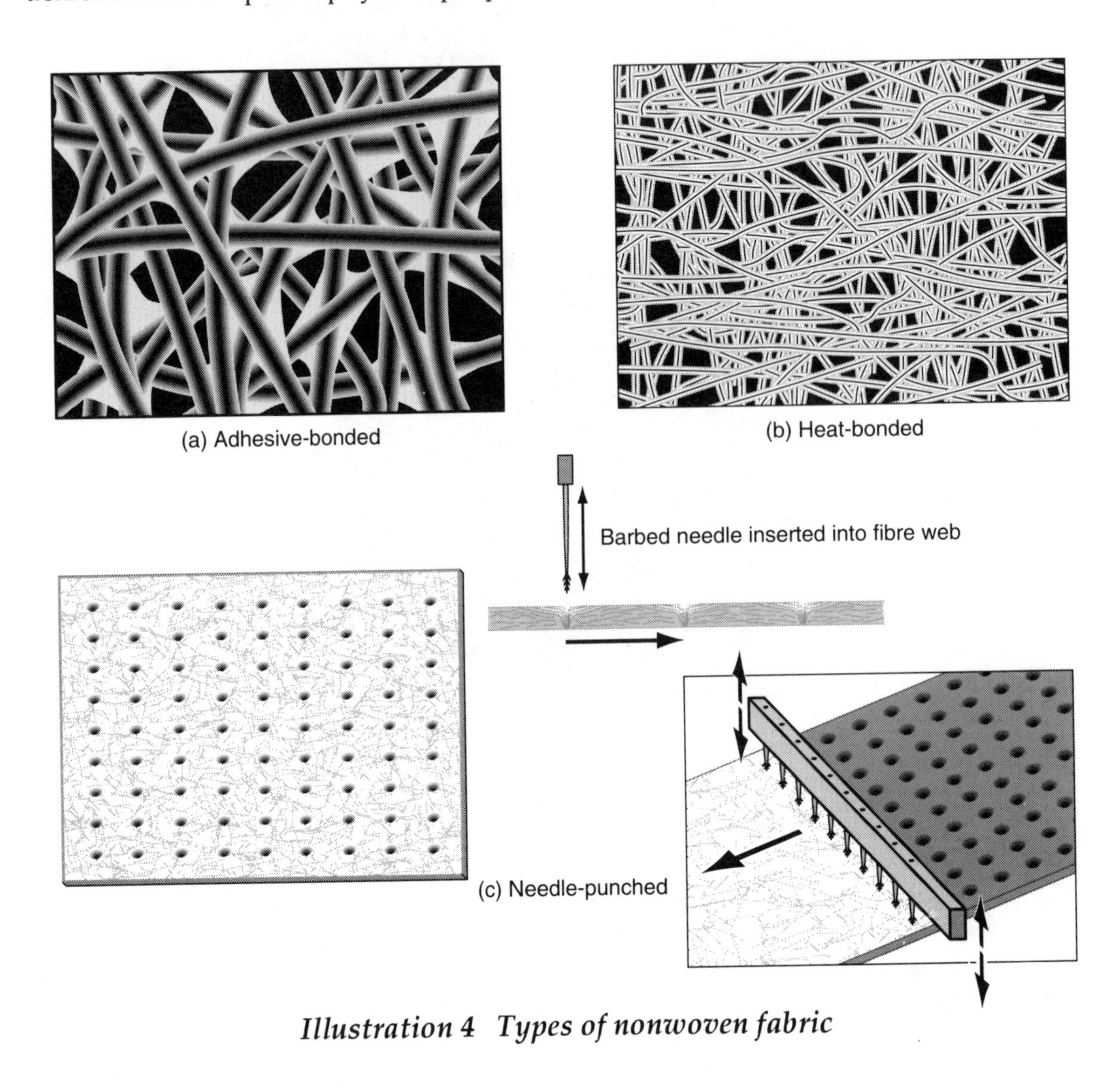

Illustration 4 Types of nonwoven fabric

Braids

Plaits, braids and sprang are all structures in which each yarn intertwines or interlaces with every other yarn at some point, but at no point does any yarn make a complete twist around another. A simple three-strand plait is a familiar structure, easily made by hand, but the intertwining of a greater number of yarns is more difficult (see Illustration 5).

Braids can be made either by hand or on high-speed equipment for the manufacture of tubes such as hosepipes or sanitary towel sleeves, cords such as shoelaces or ropes, or narrow fabrics used as garment trimmings. Sprang is a type of braiding done by hand. See Illustration 6.

Illustration 5 Braiding

Illustration 6 Sprang

Openwork structures

There are other structures that are used mainly for making open fabrics, but which can also be used for closer constructions. Most of these are hand-made, but some, notably lace fabrics, have machine-made versions. Diagrams of some of these structures are shown in Illustration 7.

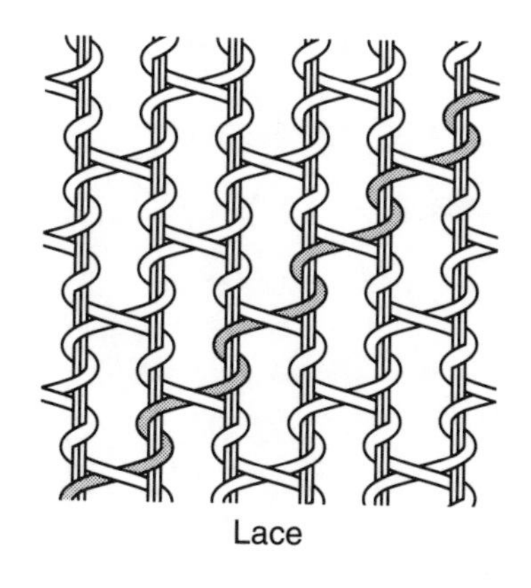

Lace

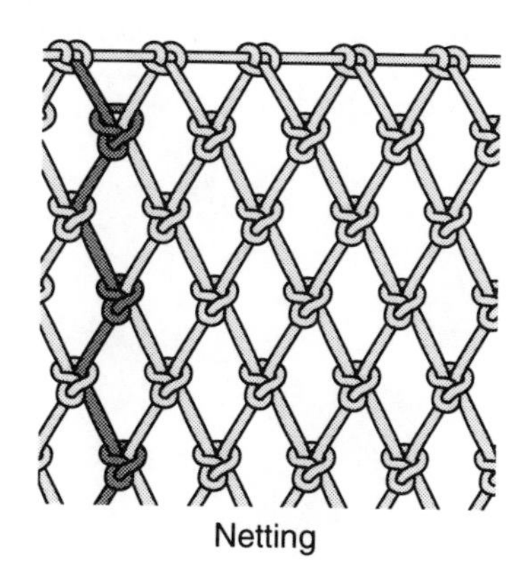

Netting

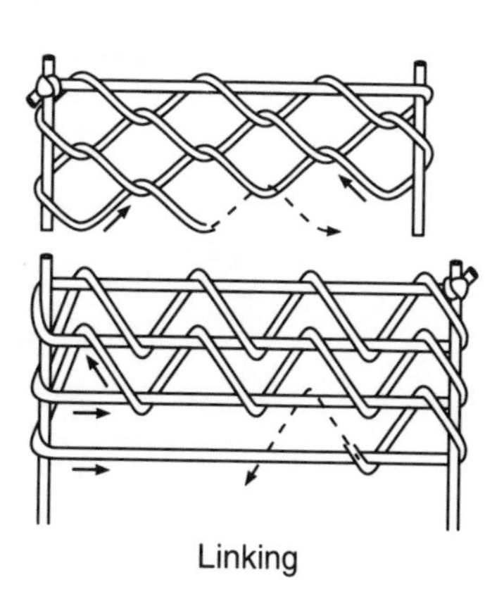

Linking

Illustration 7 Openwork structures

The half knot

The square knot

Vertical clove hitch

Horizontal clove hitch

Illustration 7a Macramé knots

Lace and weft-twining are examples of fabrics made by twisting yarns around each other. Tatting, macramé (Illustration 7a) and net-making are made by knotting, although nets made on warp-knitting machines are common in Europe and the USA. Crochet is a simple form of warp-knitting usually done by hand. Probably the simplest way of producing a fabric is by linking, in which one yarn is continually linked with itself. However, those who practice the art of linking are highly skilled and can produce beautiful structures, which are themselves intricate rather than simple.

2. BASIC PRINCIPLES AND PROCESSES OF FABRIC MANUFACTURE

WEAVING

Weaving is the process of making cloth with two components, a warp and a weft, and can be done by very simple techniques or on complicated looms. Simple looms can easily be made from basic, locally-available materials such as wood, bamboo, palm tree, angle iron or even concrete, and as such can sometimes be easily constructed on site. The process of weaving involves a warp of yarn being tied onto a loom across which weft yarn is passed; hence the loom must first of all be provided with a complete set of warp yarns arranged side by side according to the fabric width. For instance, if the width is 50cm and there are to be 10 ends per cm, then 10 x 5 = 500 ends must be arranged in the loom. The process of warping is described more fully in the ITDG handbook, *Yarn Preparation*.

In the simplest looms the warp yarns are arranged with both ends of each thread tied or fixed to two opposite sides of a rectangular wooden frame, which may be horizontal or vertical. The warp yarns need to be under equal tension. The warp yarns on some traditional vertical looms are tied to a beam above head height and weighted at the bottom by tying stones to each end. A simple vertical frame loom, shown in Illustration 8a, and a basic horizontal loom, shown in Illustration 8b, both use a continuous warp under tension.

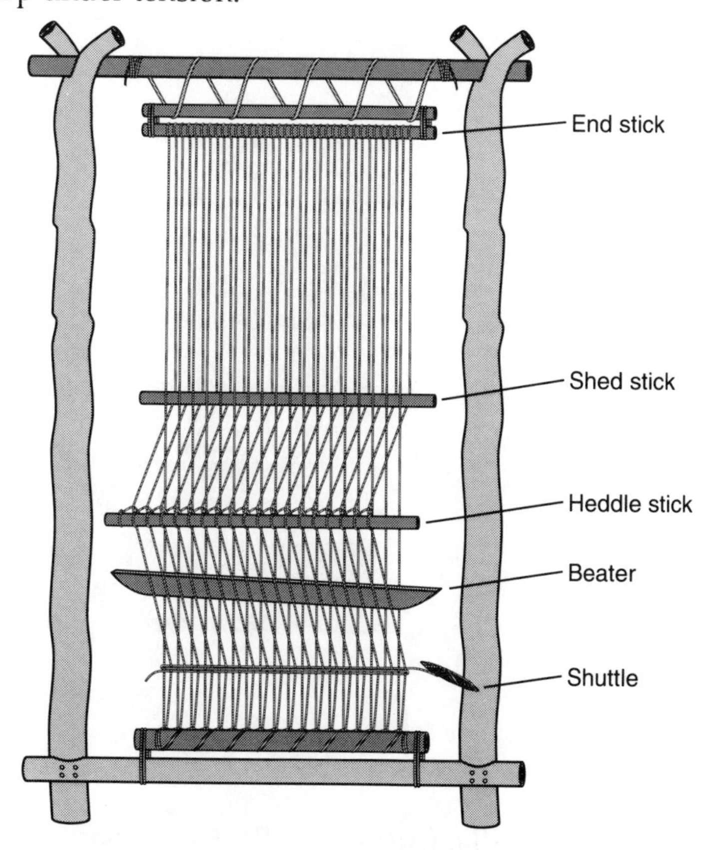

Illustration 8a Vertical frame loom

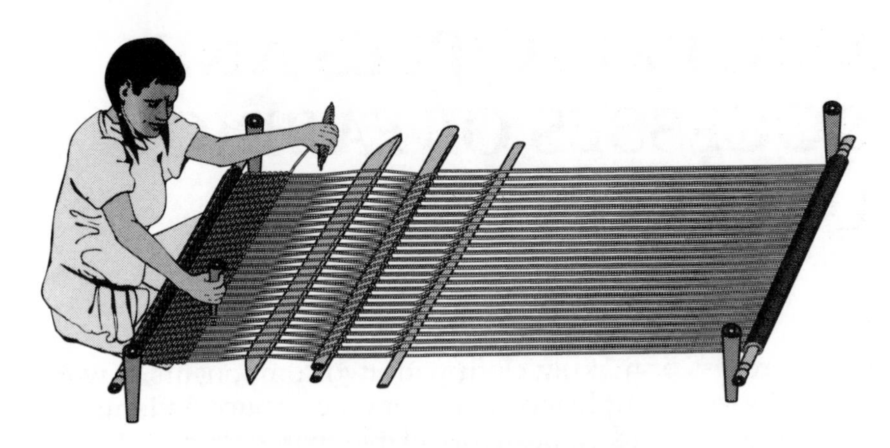

Illustration 8b Horizontal loom

These types of loom limit the length of fabric according to the size of the loom framework, and the fell of the cloth moves along the loom instead of remaining stationary. These looms may be suitable for making rugs, bed covers, wall hangings, tapestries or garment lengths of a limited size.

In order to weave a longer, continuous length of fabric, the delivery of the warp is made, under tension, from a rotating beam at the back of the loom to a take-up roller or breast beam on a handloom, on to which the woven fabric is wound, at the front of the loom. There are many variations of looms using this method and each is made according to a particular need and product; taking into account construction materials available for the loom and the raw materials for weaving. The backstrap loom, seen in Illustration 9a is one example. See also Illustration 13a and 13b. The tension of the warp is maintained by the weaver leaning back.

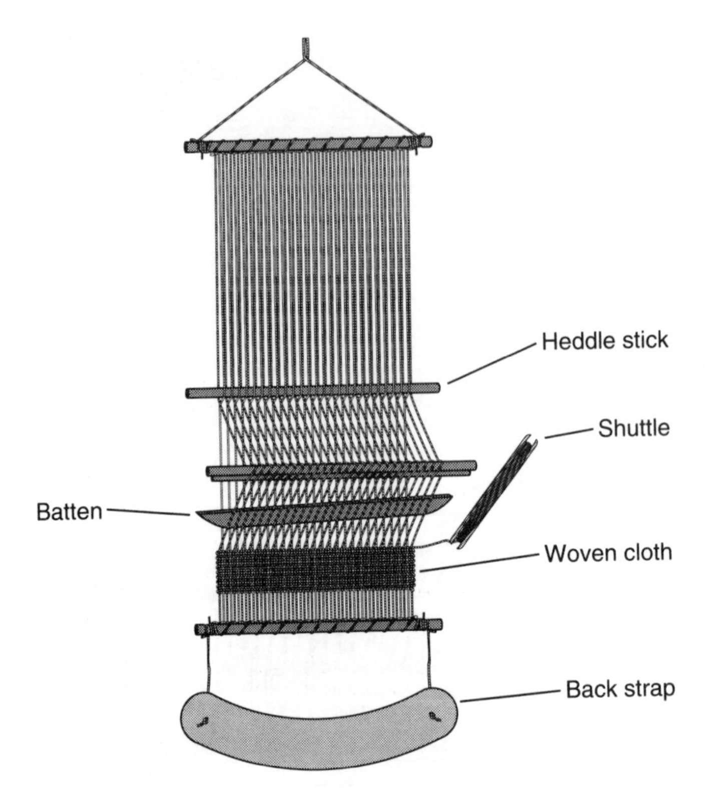

Illustration 9a Backstrap loom

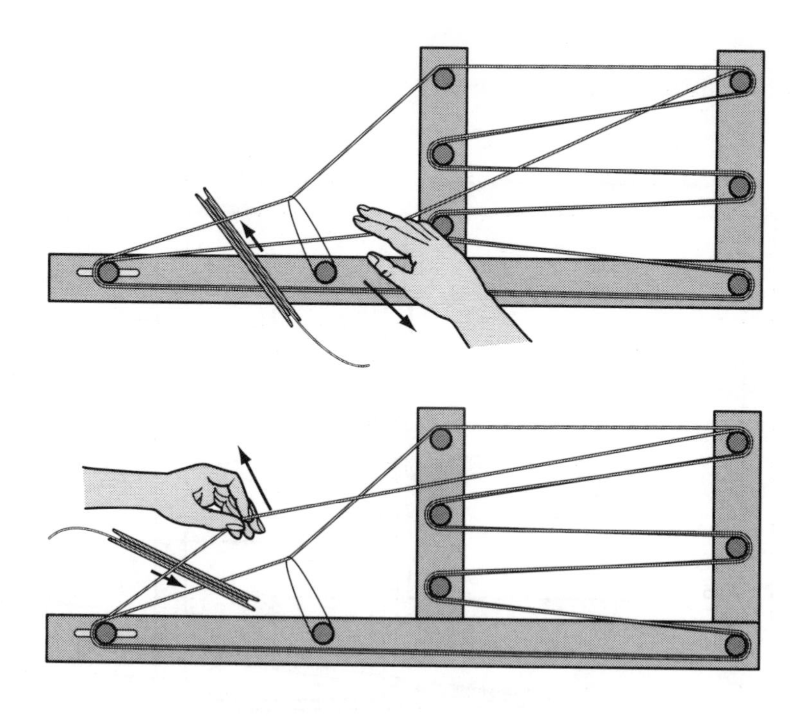

Illustration 9b Inkle loom

The inkle loom, Illustration 9b, is a simple loom for weaving narrow fabrics like belts and ribbons. Similarly, tablet weaving, using perforated cards approximately six centimetres square, is one of the simplest methods of weaving a narrow fabric without a loom (Illustration 9c).

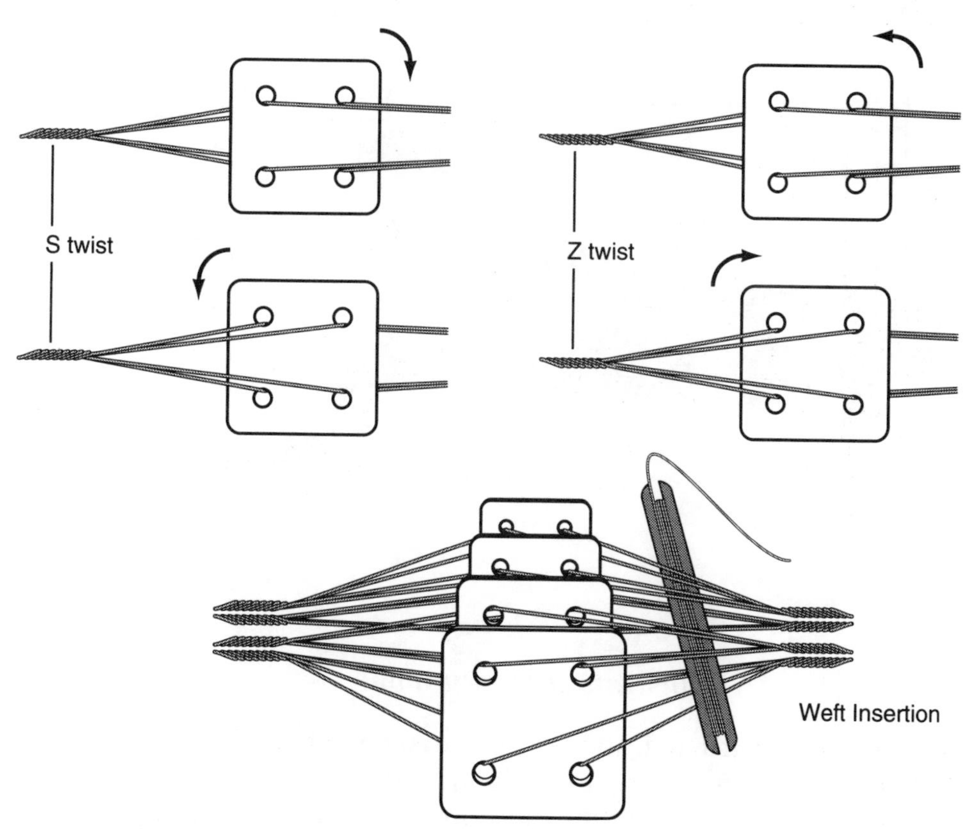

Illustration 9c Tablet weaving

The elevation of a more sophisticated wooden frame handloom is shown in Illustration 10.

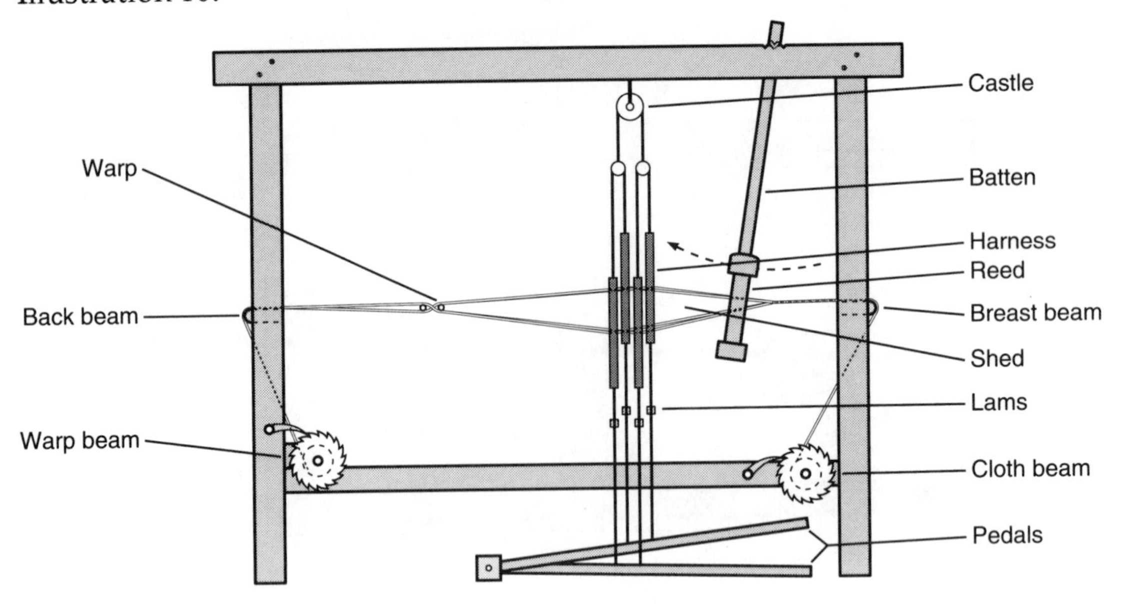

Illustration 10 Elevation diagram of counterbalance loom

A pit loom is shown in Illustration 11.

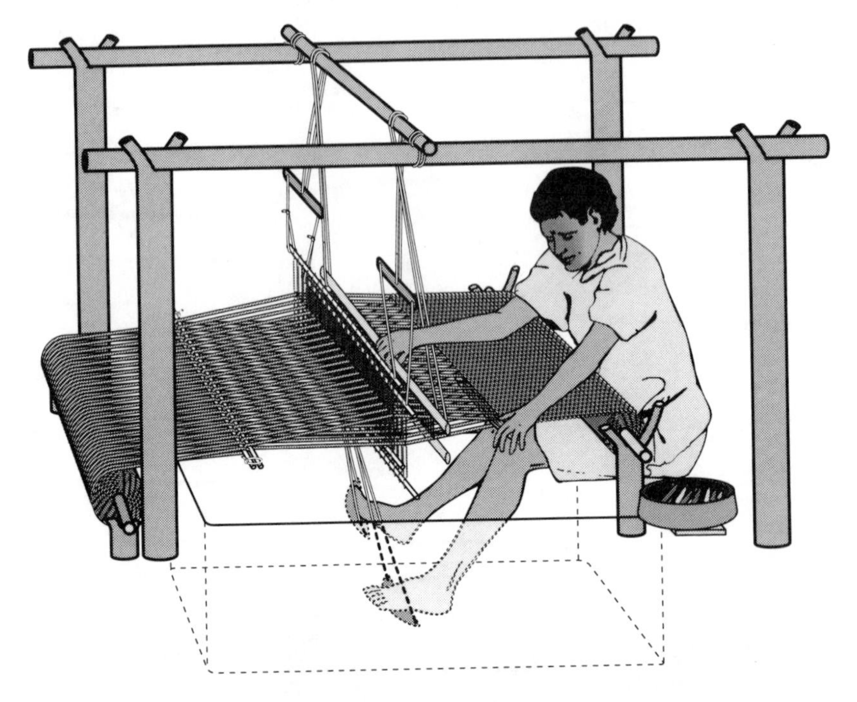

Illustration 11 Pit loom

This type of loom is constructed above a pit, dispensing with much of the substructure of the loom seen in the previous illustration, so that the weaver sits on the edge of the pit with the front, take-up beam immediately in front of him and just above his knees. The weaver controls the pedals in the pit.

There are many variations of the weaving loom, and usually the more complicated or sophisticated the loom, the faster the fabric production, or the more intricate the fabric structure.

The weaving sequence is a repetition of the following operations:

- ☐ shedding
- ☐ picking
- ☐ beating up

These operations are performed on all looms, whether hand- or power-operated.

Shedding

Shedding is the creation of an opening for the weft by lifting some of the ends and leaving the remainder lowered. It is possible to thread the weft over and under the warp ends one by one, but it is better to have a device which will lift or lower the threads. For instance, to make a plain weave fabric all the odd ends are lowered and the even raised for one pick insertion, while for the next pick the reverse happens. The most simple, mechanical means of shedding is with a fixed-heddle reed which combines warp spacing, order and lifting with the reed.

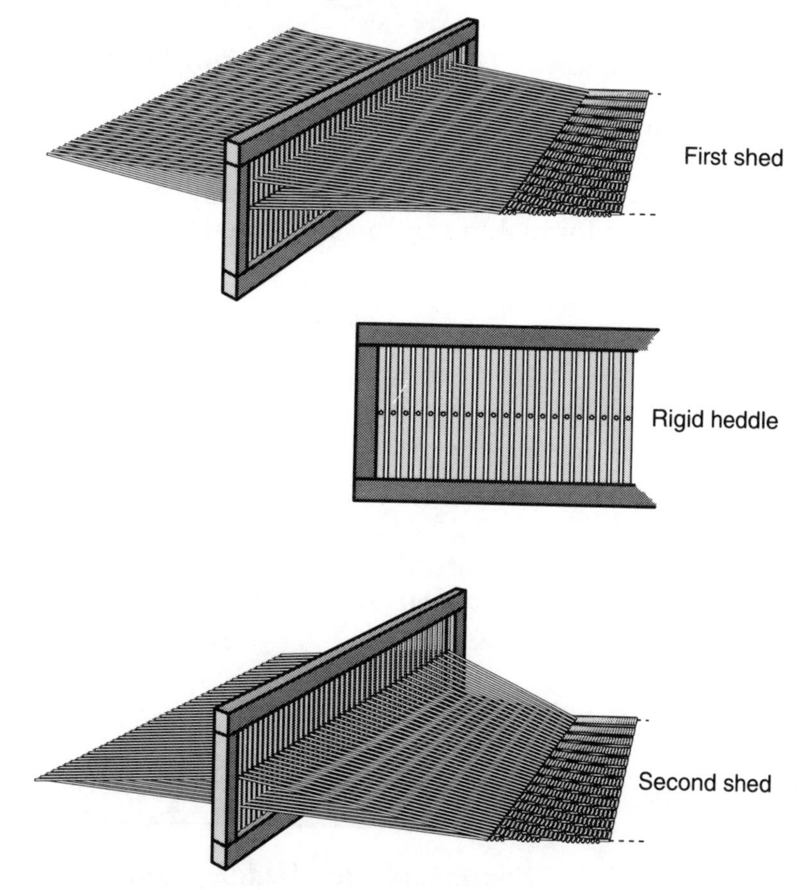

Illustration 12 Rigid or fixed heddle reed showing alternate shedding

Illustrations 12 shows the rigid or fixed heddle reed raising and depressing the alternate warp ends. The method of shedding on the backstrap loom is shown in Illustrations 13a and 13b, in which a shed-stick is used to cause the odd ends to be raised, creating an opening for the weft; the even ends are all lifted simultaneously by the heddle stick on the next pick.

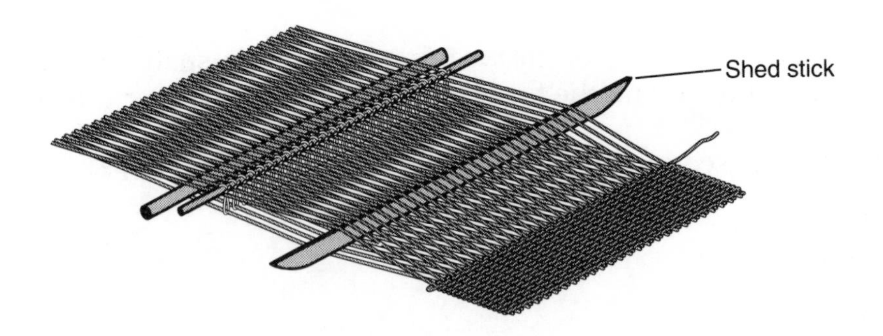

Illustration 13a First shed on backstrap loom using shed stick

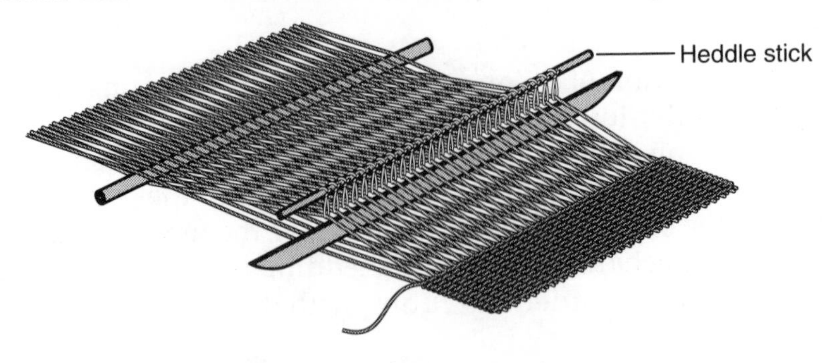

Illustration 13b Second shed using heddle stick

Normally warp yarns are threaded individually through healds which are attached in order onto shafts (healds are sometimes called heddles, and shafts also known as frames or harnesses). Illustration 13c shows three types of heddle. Illustration 13d shows a heddle board for making cotton string heddles.

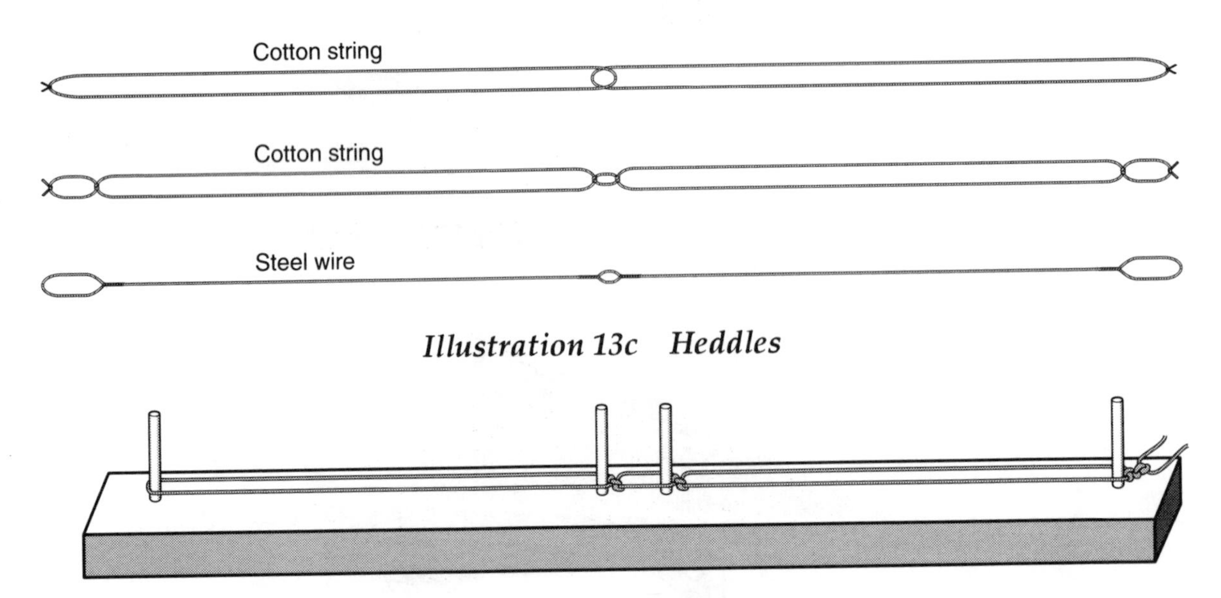

Illustration 13c Heddles

Illustration 13d Heddle board

The shafts are raised and lowered to control the lifting of the warp threads. The pit loom in Illustration 11 has two shafts. For more complicated weaves, more shafts are required, such as the four shafts in the frame loom in Illustration 10. Looms such as the counterbalance loom have a system of pedals and pullies to control up to 4 shafts (see Illustrations 10, 14a and 14b).

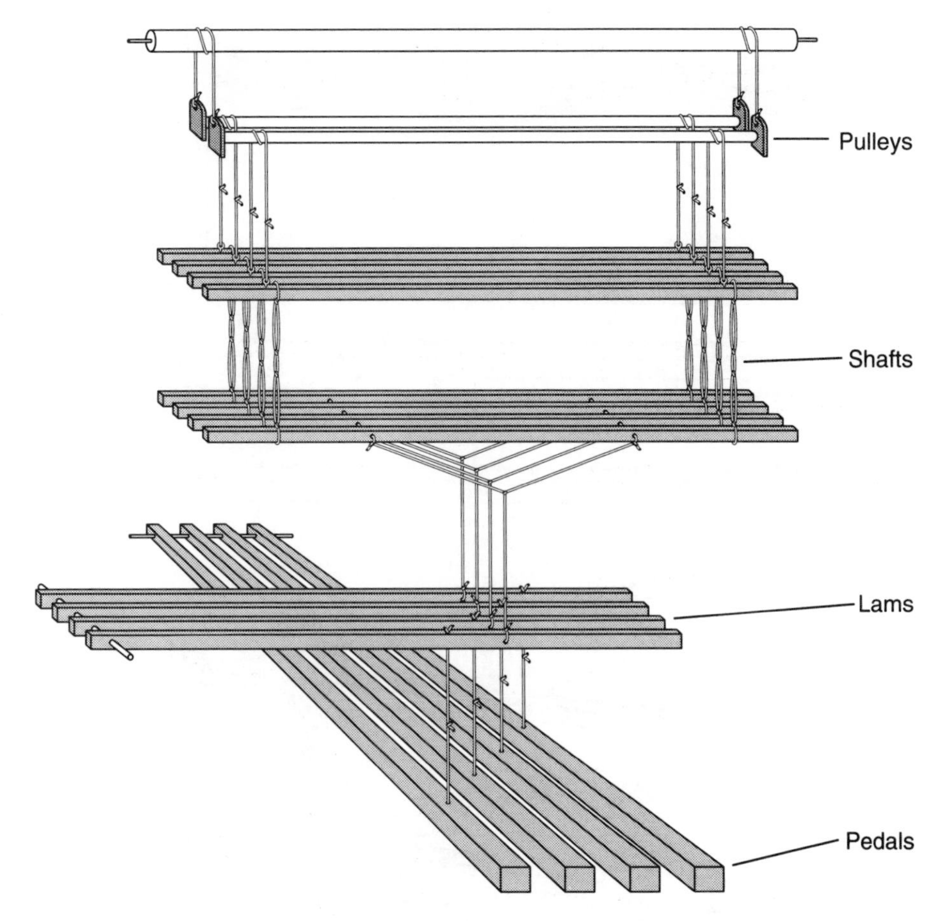

Illustration 14a Counterbalance system

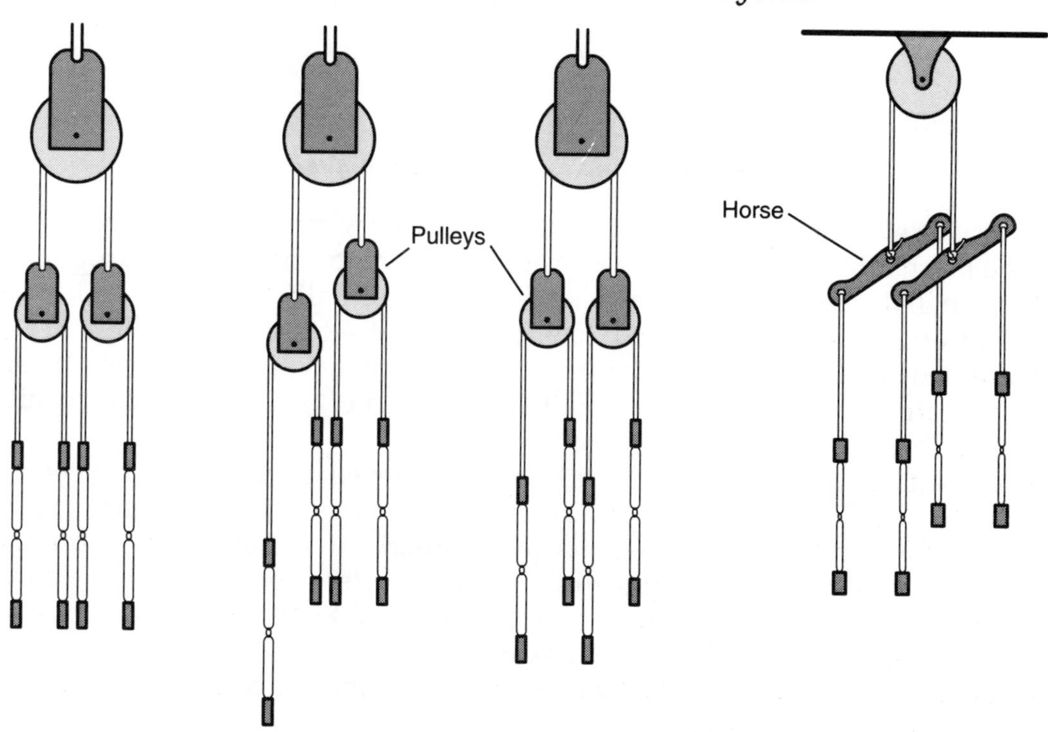

Illustration 14b Counterbalance motions

The countermarch system can control up to 12 shafts and allows even more complex weaves to be produced (see Illustration 14c).

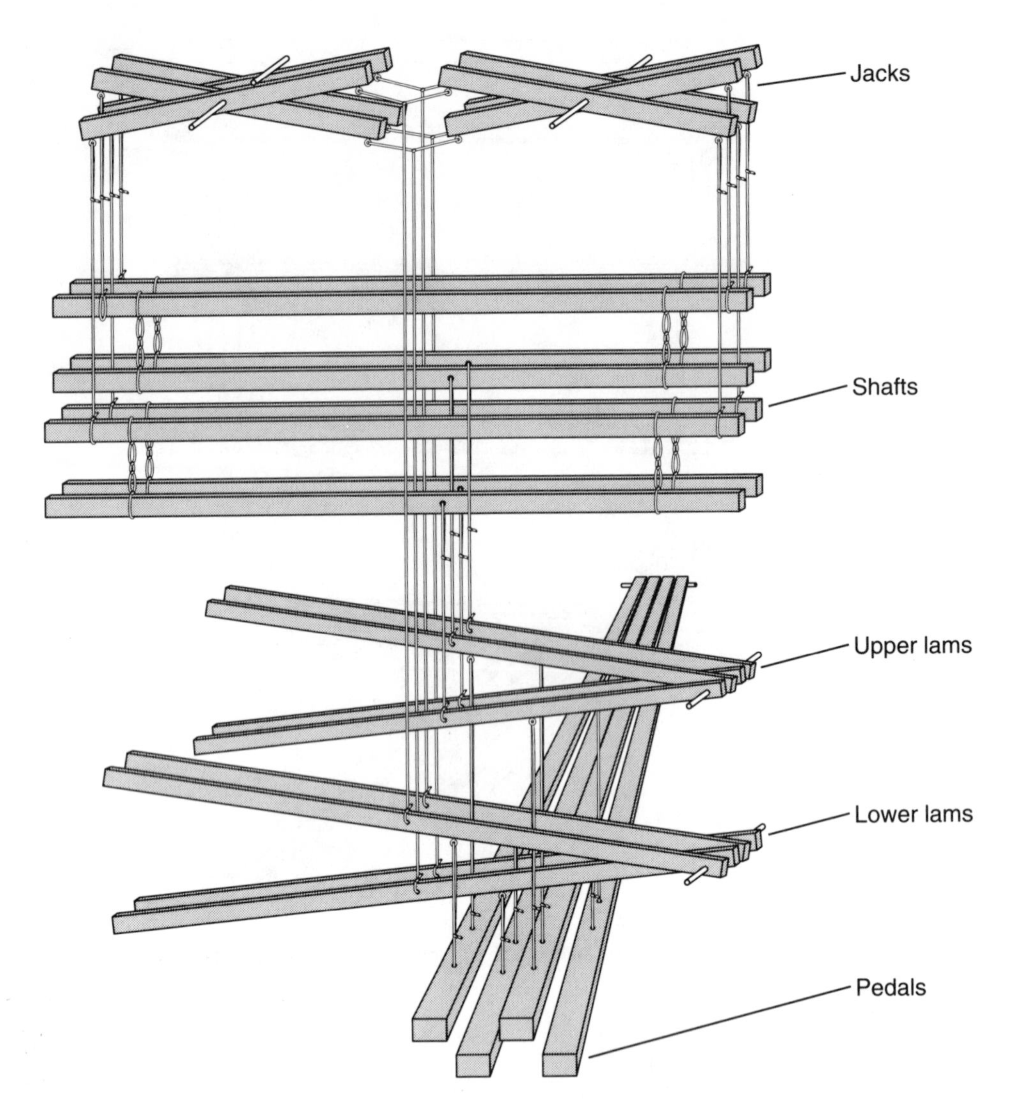

Illustration 14c Countermarch system

To control an even larger number of shafts, a dobby or witch mechanism can be used. This is a mechanism for controlling any number of shafts and it can be programmed to lift them in the required sequence. The dobby is normally fixed above a loom and one shedding lever only is used to lift the required shafts. Dobbies are pre-programmed; the simpler ones with chains of wooden lags and pegs or small punched cards. More recently developed dobby mechanisms are electronically controlled.

The most complex system of shedding is achieved by a jacquard mechanism attached to a loom. This system, invented by Frenchman Joseph Marie Jacquard between 1801 and 1810, allows large numbers of healds to be controlled individually, enabling large, sophisticated weave patterns or designs to be created. Jacquard mechanisms are made to control a range of different numbers of individual ends from 100, 200, 400, 600 etc. to a maximum of 2688.

Jacquard mechanisms are controlled either by punched cards or electronically. They are costly to buy, maintain and use.

Picking

Picking is the insertion of the weft across the loom through the open shed. The simplest way to do this is to wrap a length of weft yarn around a stick, as shown in Illustration 15.

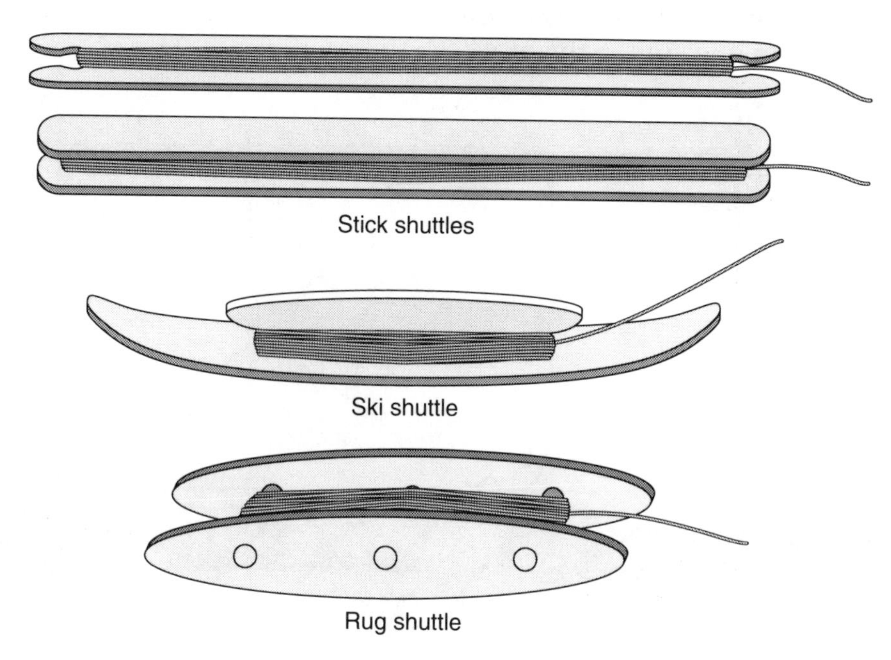

Illustration 15

The stick shuttle is passed through the warp from side to side by hand, the weft being gradually unwound. Other types of shuttle are shown in Illustrations 16 where the yarn is wound on a bobbin, pirn or quill which is placed in the shuttle.

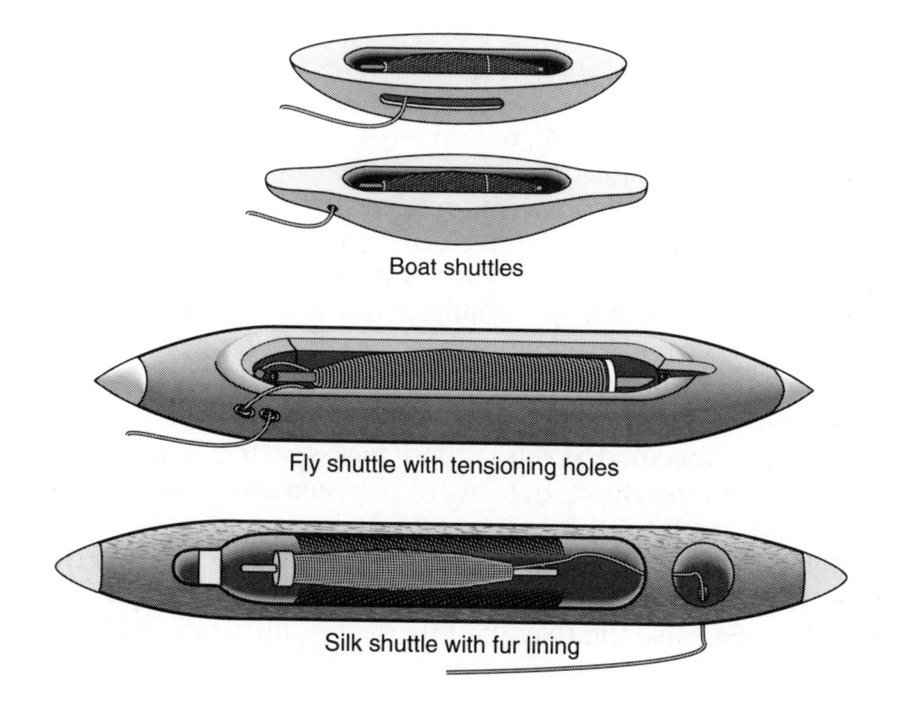

Illustration 16

The pirn continually unwinds as the shuttle travels through the shed, across the loom. Normally, if the fabric width is greater than 85cm it is difficult for a weaver to throw and catch a shuttle from side to side for any length of time. Improvements to the design of the loom were made so that the weaver could propel the shuttle, by a simple mechanism, across a much wider cloth. The fly-shuttle was invented in the 18th century, (see Illustration 17), which allows the weaver to use a handle to propel the shuttle from one side to the other.

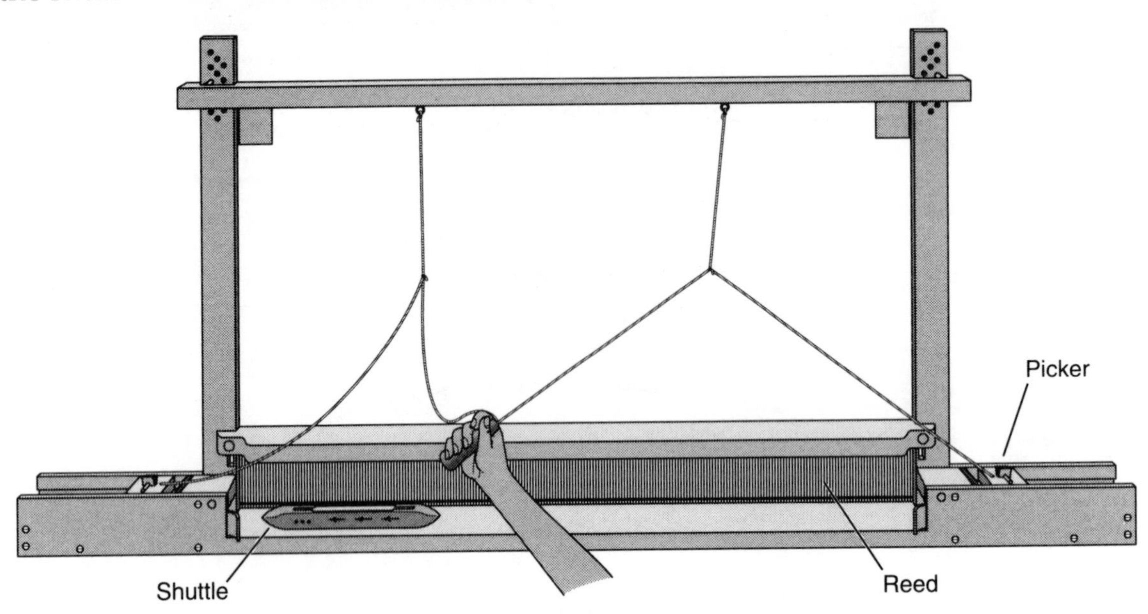

Illustration 17 Fly shuttle operation

The fly-shuttle technology is the basis of all power shuttle looms, in which the shuttle is propelled from side to side by being hit with a picking stick.

Beating up

Beating up is the action of pushing the newly-inserted pick closer to the fabric, sometimes known as the fell of the cloth, on the loom. On simple looms a long piece of wood is used, known as a 'sword', but on most looms the pick is beaten up by the reed fixed into the sley. In handloom weaving, typically, one hand is used to insert the pick, and the other used to pull the reed forward, causing the picks to pack together. Variation in the pressure exerted during beating up may cause the spacing from pick to pick within the fabric to vary, hence the weaver needs to beat up consistently in order to weave good quality cloth with the required specification. During the process of weaving therefore, beating up is an important operation.

On power looms, and handlooms where the movement of the reed is controlled mechanically in conjunction with the other movements on the loom, the reed always beats up to the same specified position, and pick-spacing is determined by how much the front roller moves the cloth forward between pick insertions, so a constant pick-spacing should result.

The reed determines the arrangement of the warp threads across the fabric, and the action of beating up therefore controls the positions of the warp and weft threads in the woven fabric. See also the fixed heddle reed in Illustration 12.

High production looms

The so-called efficiency of the weaving operation depends upon looms running continuously for long periods of time without stopping. On a normal shuttle handloom, semi-automatic loom, or basic power loom, the pirn in the shuttle contains a limited amount of weft yarn which must be replaced at frequent intervals, causing a break in the flow of weaving. This situation can be improved by the use of more than one shuttle box at the side, allowing a quick change from one shuttle to another; this also allows the use of different colours of weft. The more sophisticated power loom is equipped to replace empty pirns automatically with full ones, without stopping the loom.

Developments in power loom manufacture have resulted in shuttleless looms. There are several types of these, but they all cause the weft to be taken across the loom from a single, large, stationary package at the side of the loom. They are fast and efficient, but capital-intensive technologies, generally suitable for large-scale production. They do not have the flexibility of cheaper power looms or handlooms, which can weave a variety of different structures. The four main types are as follows:

Gripper shuttle loom. The gripper shuttle loom, known as the Sulzer loom after the Swiss company who developed it, has a small metal gripper which picks up the weft yarn and projects it across to the other side.

Air-jet loom. The air-jet loom uses a jet of air under high pressure to blow the weft yarn across the warp.

Water-jet loom. The water-jet loom uses a fine, high-pressure jet of water to propel the weft yarn across the warp.

Rapier loom. The rapier loom has a moving arm or rapier which picks up the free end of weft yarn and takes it across, either to the other side, or half-way across, where a second rapier collects it and transports it the remainder of the way.

In all of these methods of weft projection, a single large package (or several if different colours of weft are required) stands at the side of the loom. The free end is brought to the point of entry to the warp shed, the correct length is measured from the package for one pick, the pick is passed across the loom and the yarn is cut, prior to the whole operation being repeated.

KNITTING

Weft knitting

The knitted structure is made by interlocking loops of yarn, formed by taking the free end around one needle and transferring it to another. In hand-knitting two long needles, or pins, are used, although more are sometimes used to make a tubular fabric structure. Needles are available in a range of thicknesses. They are basically two sticks, their thickness controlling the size of stitch. For heavier fabrics, when a coarser yarn is used, creating a larger stitch size, thicker needles are required. Knitting proceeds across the width of the fabric, the total width being governed by the number of stitches in one row or course.

Illustration 18 overleaf shows the four stages of making a stitch in hand knitting, usually labelled 'in, over, through, off,' indicating the insertion of the needle into a stitch on the previous course, creation of a new loop, passing the new loop through the previous one, and clearing the previous one from the needle.

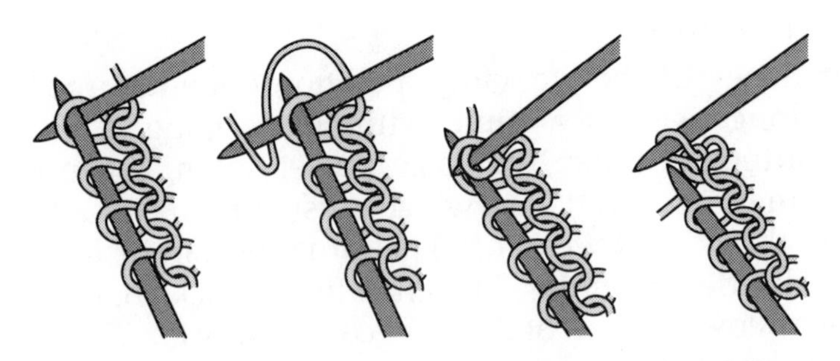

Illustration 18 Hand knitting with needles

Variation in these basic actions can produce a variety of structures. The possibility of altering the number of stitches across, and therefore varying the width of fabric, allows garment shapes to be produced without the waste which occurs if garments are made by cutting shapes from continuous lengths of fabric.

There is scope for a knitter to design each garment individually. Hand knitting is slow to produce, even for a very experienced knitter.

Machine weft knitting

The knitting machine consists of a needle bed containing a large number of needles, one for each wale, or column of stitches, of the fabric. One course is knitted by moving a carriage across the needle bed in one operation. The speed of machine knitting is considerably faster than hand-knitting. The hand-operated knitting machine and the motor-operated knitting machine work on the same fundamental principles. The carriage which delivers yarn to the needles also contains a cam system which controls the operation of the needles. There are several types of needle, but in the vast majority of weft knitting machines the latch needle is used. The operation of this needle is shown below.

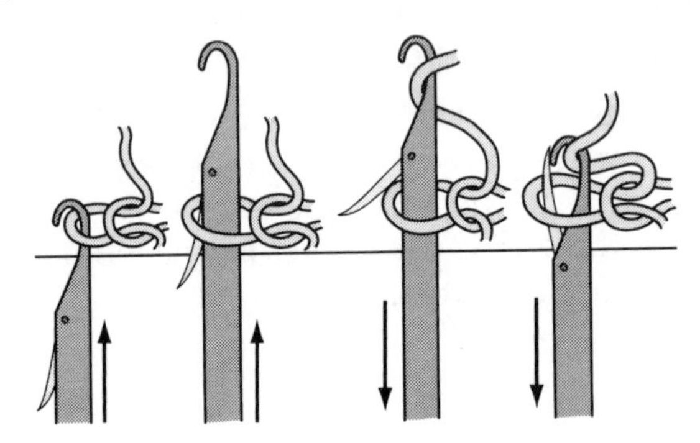

Illustration 19 Operation of latch needle on knitting machine

The needle movement is vertical, controlled by the movement of the cam as the carriage passes across the needle bed. The stitch is formed as follows. The needle rises to allow yarn to enter the needle head; it then drops, pulling the yarn into a loop and at the same time the previous loop slips over the needle latch and clears the needle. The stitch cam on the carriage can be adjusted to set the distance the needle moves down, and this operation controls the length of the stitch.

Knitting machines are available with many additions to the basic machine, increasing the patterning potential. The flexibility of producing a wide variety of fabric weights is not possible on a knitting machine as it is with a loom. A knitting machine will contain a needle bed which has a fixed size of needle and a fixed distance between the needles. The thickness of the yarn and the stitch length which can be used are therefore limited. The number of needles per inch is called the 'gauge'. A fine-gauge machine is one with a large number of needles across the width, a small distance between each, and in which the needles themselves will be thin: made to knit fine yarn. Coarse yarn can not be knitted on this type of machine. On the other hand, it is not normally feasible to knit fine yarns on a coarse-gauge machine, which would produce a very open structure. Weft knitting is almost exclusively used for the production of garments, and fabric suitable for being made into garments. If fine yarns are used with large stitches, limp fabrics are produced. If coarse yarns are used with low stitch-lengths, stiff, cardboard-like structures result. A knitting machine will therefore be used for a small range of fabric weights and end-uses.

Relationship between machine gauge and yarn counts

Gauge (needles per inch)	*Yarn counts*		*Approx weight knitted (g/m^2)*	*Type of Garment/fabric*
	TEX	*Worsted*		
2.5	400-900	1-2s	300	heavy sweater
3.5	300-600	1.5-3s		
5	120-300	3-7s		
7	110-150	6-8s	250	medium sweater
8	80-125	7-11s		
10	60-80	11-15s	200	skirt/trouser
12	40-70	13-21s		
14	35-55	16-26s	150	dress/T-shirt

Classification of weft knitting machines

Weft knitting machines can be either flat or circular. A circular machine has the needle bed arranged in a complete circle, so that the yarn knits in a spiral, producing a tube of fabric. Circular machines are widely used industrially for fabric production; the tube of fabric is cut lengthwise to produce wide, continuous lengths. The machines normally knit from a large number of yarn packages simultaneously so that on one revolution a number of courses are produced. Some circular machines are used as garment-length machines; these are mainly used to make socks and stockings. A hand-operated circular stocking or sock knitting machine is shown in Illustration 20.

Flat machines, in which there are one or two beds of needles in straight lines running parallel to each other, are the chief type of hand-operated flat machine. The simplest machine allows a wide range of fabric structures and patterns, depending on the skill of the user. It is a fast way of making heavy fabrics. Light fabrics, having a higher number of stitches, are difficult to make to a high quality by hand. One advantage of knitting over other types of fabric production is that only one yarn is needed, which minimizes yarn preparation and makes the technology more portable.

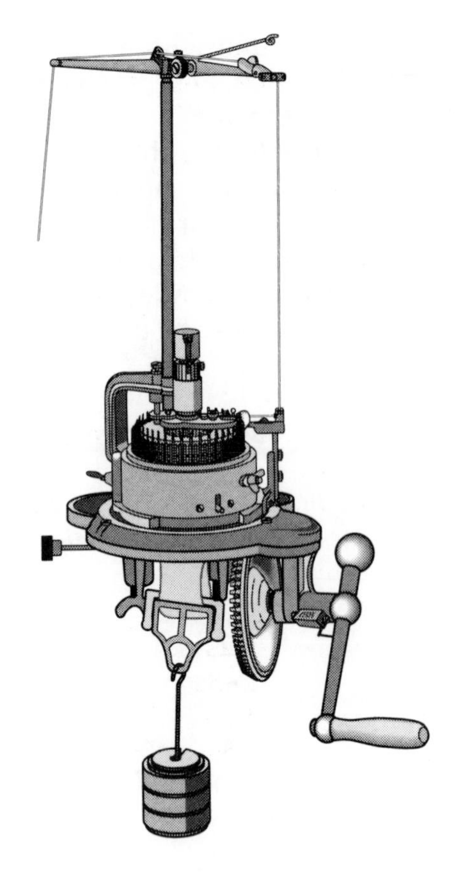

Illustration 20 Hand-operated circular stocking machine

Fabric machines

Either flat-bed or circular machines are used for the manufacture of continuous lengths of fabric which can then be cut into garment shapes.

Fully fashioned garment machines. These are used to weft-knit individual pieces of a garment. For instance, by varying the number of wales across the garment from course to course, narrowing and widening can be achieved as the garment piece is being knitted and shaped. The pieces (eg. arms, front and back) are then sewn together to produce a garment. The term 'fashioning' refers to the technique which allows the garment pieces to be knitted without the waste which is otherwise caused by cutting garment shapes from a length of knitted fabric. Fashioning is slower and therefore less productive.

Garment length machines. Fabrics with large pattern areas made with several colours of yarn can be achieved by hand-knitting on two needles or pins. On fine gauge circular or flat-bed knitting machines, extra pattern mechanisms need to be introduced, which increases the capital cost and the cost of maintenance. However, there are domestic machines which can satisfy most of the requirements of the knitter. Illustration 21 shows a small, robust and lightweight v-bed knitting machine.

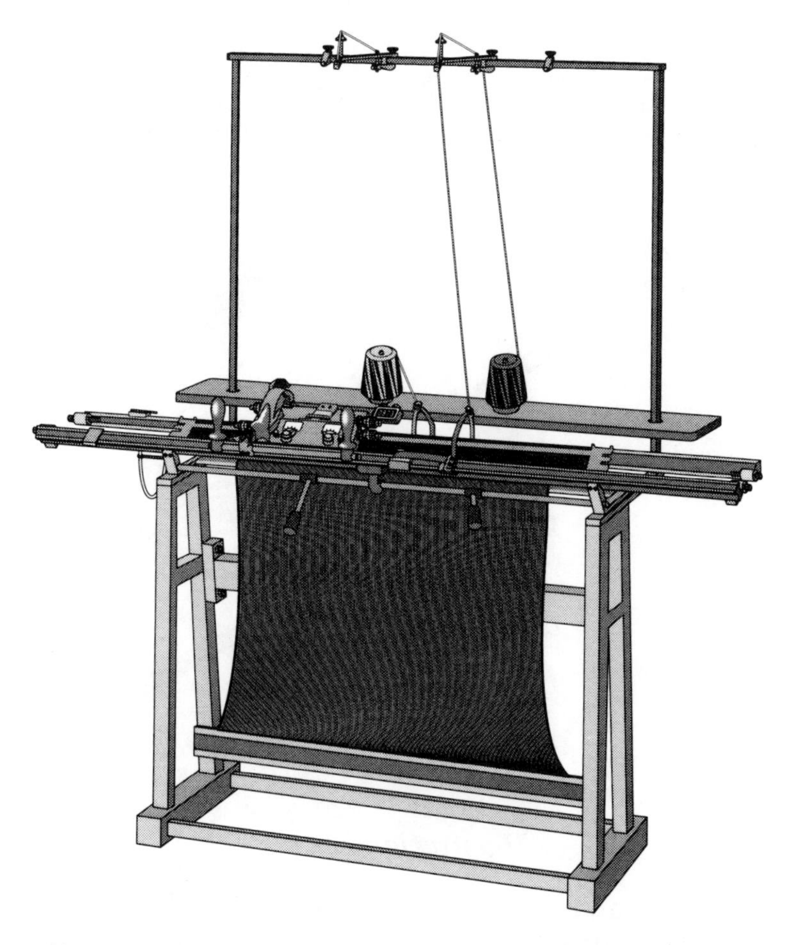

Illustration 21 V-bed hand-operated knitting machine

A difficult problem for an isolated machine knitter is that of needle breakages, which occur with increasing frequency as the gauge of the machine increases. This type of coarse gauge machine overcomes this difficulty to a large extent as it has heavy strong needles that are less likely to break.

Warp knitting

The warp knitting machine is one which uses a set of yarns, normally fed from beams similar to weaving beams, and delivered to a needle bed similar to that on a weft knitting machine. Fabric is created by the formation of knitted loops; the overall direction of the yarns is down the fabric length. If each yarn were continually fed to the same needle, then a series of discontinuous chains would result, (see the paragraph on crochet below), but the yarns are controlled by guide bars which can deliver the yarns to different needles on different courses, and the stitches are therefore connected together to form fabric. Warp-knit fabric production is aimed at markets which are either specialized, or high-volume and low-cost. The machinery is expensive and made for very high production speeds, such as 2000 courses per minute. In addition, the machines only operate efficiently with high quality, fine, synthetic yarns so the raw material cost is also high. Hi-tech warp knitting is therefore of limited direct interest to the small producer.

CROCHET

In hand crochet, a single hooked needle is used to pull loops of yarn through previously-made loops to create chain stitches. Illustration 22a shows how a continuous chain can be created, starting from a single loop.

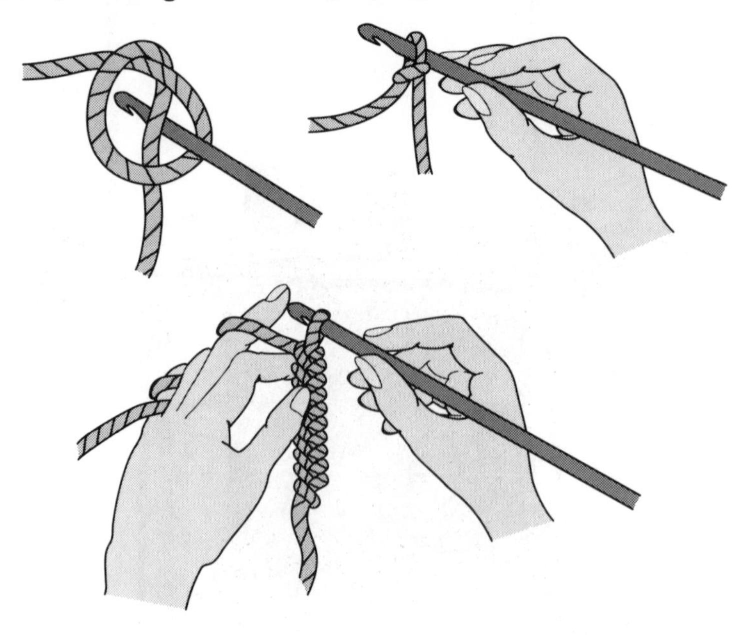

Illustration 22a

The yarn is held in one hand and the hook in the other. Basic fabric is made by forming a row of stitches with a length equal to the fabric width and then returning along the chain making double crochet stitches by inserting the needle into previous loops to make a second row of loops joined to the first, see Illustration 22b.

Illustration 22b

The technique can create solid fabric or open lace-like fabric, and only requires a ball of yarn and a single needle. Like hand knitting, many patterns are available for crocheting garments, and the technique can be acquired with very little training. Its main disadvantage is low speed of production.

Machine crochet

The crochet machine is a simplified form of warp knitting machine. The knitting action forms continuous chains as in hand crochet, but is also has guide bars which lay thicker yarns across the knitted stitches, and so the fabric has the properties of a coarse-gauge warp-knitting machine rather than those of a hand crocheted garment; it does not stretch like a weft-knit fabric.

The technique is simpler than that of conventional tricot or raschel warp-knitting and can produce interesting fabrics with coarse yarns, so therefore is the most attractive to the small producer.

BRAIDING

Machine braiding is a somewhat complicated process for producing what are frequently no more than thick cords. Most braiding machines are used to create tubular structures such as hosepipes or cords, ropes or shoe laces. The braid is created from a number of interlacing yarns. At any one time half the yarns are travelling in one direction at some angle to the axis down the fabric, while the other half are travelling in an opposite direction, passing over and under the strands of the first group. This requires individual control of each yarn, which in turn means that the greater the number of ends, the greater the size of the machine. The simplest form of braiding is the plaiting of three strands.

Illustration 23

A multi-end hand-made braid is shown in illustration 23 where the direction of the yarns are reversed and are held in place by the insertion of a stick in the middle. The fabric is limited in length by the length of the framework itself; it is considerably shorter than the yarn length because of the angle at which all the threads lie to the axis. The operator must continually shorten the distance between the ends to allow for the take-up.

Braiding machines work like a traditional maypole (see Illustration 24).

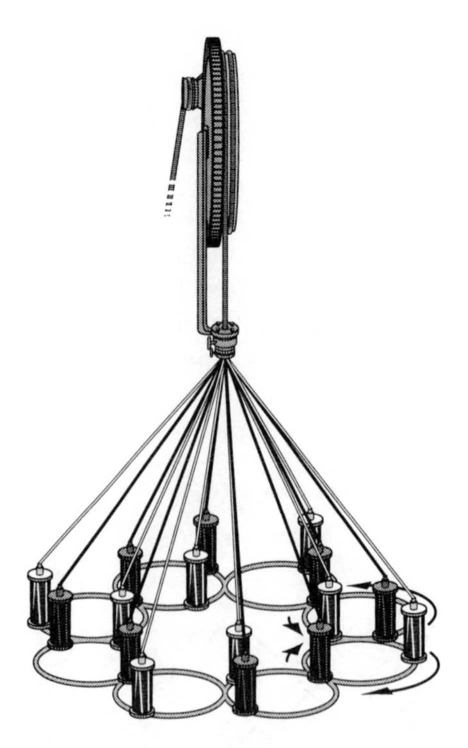

Illustration 24

Each yarn is wound on a spool or bobbin which moves horizontally around the machine in a paths like those in Illustration 25. The braid will be tubular.

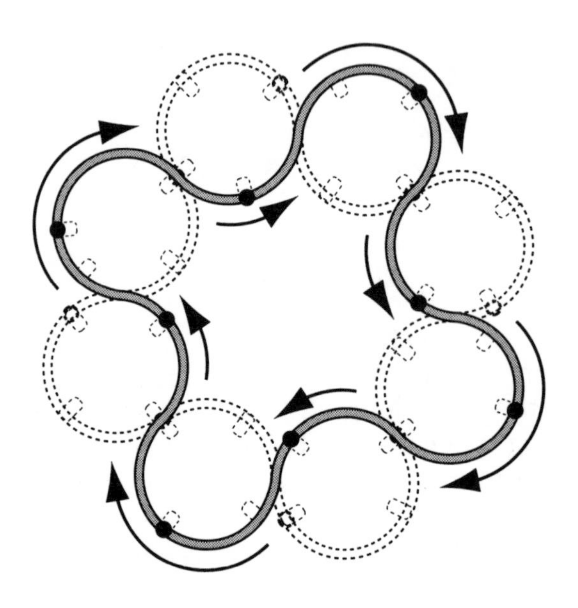

Illustration 25 Braiding path for tubular braid

The complication of the path of the carrier holding each yarn, and the variation in yarn tension as the distance of the yarn from the axis changes, means that the modern braiding machine is highly engineered. There is not as yet a common piece of equipment for making flat fabric, other than very narrow trimmings. The most common use for the braiding machine is to produce cords, and hence it is more commonly a yarn machine than a fabric machine. When used as a fabric machine, it creates fabrics with the yarns lying at an angle to the axis.

FELTING AND NONWOVENS

Felting

Felting is the process of making a fabric from animal fibres without first making a yarn. A felted fabric is formed by carding animal fibres into a loose mass or web, then subjecting them to a constant mechanical action in warm wet conditions. The fine scales on the surface of the fibres cause them to move about in all directions and gradually become tightly entangled. If the web was originally formed into a flat sheet, and flattened again after the fibres have been entangled, a strong dense fabric is formed.

To make thicker or heavier felted fabric, the original web must also be thicker. But making very thin, light-weight felts can be very difficult.

If the felting action is prolonged, the mass of fibres can become very dense and hard. The resulting felt could be used in the manufacture of filters, gaskets and washers, shoes, bags or papermaking felts.

The process of felt-making is simple. The mechanical action of pounding or kneeding the felt web is done by hand, walking on the web, rolling the web between rush mats or squeezing the web between rollers. Making very large pieces of felt by these methods is difficult.

All animal fibres will felt together, but finer fibres will felt more easily than coarse. Sometimes felting can be a disadvantage, particularly during the scouring, washing and dyeing of yarns or fabrics, when they may felt by accident.

Woven or knitted fabrics can be felted, obscuring the structure of the fabric, and can undergo a similar process as felting loose fibres. Although this process is loosely termed 'felting' the more precise designations are 'milling' or 'fulling'. Berets are often manufactured using this technique.

A sophisticated process of producing heavy yarns by felting for use in carpets has been developed in recent years.

Some plants have natural interlacing structures, usually the inner bark or 'bast', and with soaking, softening and flattening a smooth, felt-like fabric is made. Bark and Tapa cloths from the Pacific Islands and Papyrus from the upper Nile are well-known examples.

Nonwovens

Nonwovens are produced from carded webs of natural or man-made fibres. In general, they use high capital-cost equipment, produce fabrics at high speed and therefore require a high throughput of fibre. Nonwovens can also be produced on small-scale equipment or even by hand. It is possible, for example, to soak small samples of web in a polymer emulsion such as natural rubber latex, and then use some form of heat to dry the fabric. This would create an adhesive-bonded fabric, which could be used as an interlining or as an absorbent wipe or for sanitary use. The main problem is that large-scale industrial concerns can produce these basic fabrics extremely cheaply and quickly, and there is little scope for small scale production.

3. SIMPLE METHODS OF DETERMINING THE QUALITY OF FABRICS

QUALITY CONTROL

The production of fabric often involves a number of people, and the working standards of each person is affected by the others. This overlapping and interdependence of tradespeople and craftspeople involved in fabric production makes quality control a difficult aspect of small-scale textile production. Problems may arise during weaving which originated during spinning (see ITDG handbook, *Spinning*), or during warp preparation (see ITDG handbook, *Yarn Preparation*). It is therefore of prime importance to locate the point in the sequence of production which is causing problems. Uneven, poor quality yarn, patchy dyeing or an excess of knots in the yarn are not within the control of the fabric maker, but the fabric will be adversely affected by them.

Within a small-scale fabric manufacturing enterprise there are a number of ways in which the quality of fabric can be affected and controlled. The environment within which production takes place is important. If people work in an adequate space and reasonable temperature, for instance, the problems of quality control will be reduced. Similarly, if they have good tools and raw materials, poor quality fabrics will be less likely to be produced.

Overlaying both the circumstances of production and the control of the process is the organization of payment for the worker's services. If a worker is paid by the hour, production potential may not be reached. If on the other hand payment is by the metre, known as 'piece-work', then the quality may suffer. A balance must be struck between these two extremes and a system evolved which ensures the maximum production possible within the constraints imposed by quality control restrictions and the health and safety of workers.

Woven fabrics

During the weaving process itself two aspects are critical, the beat up and the selvedge.

Beat up, which controls the weft sett (the number of picks per cm in the fabric) is controlled by the swing of the reed (the batten). Regular operation of the reed is a matter of practice and concentration. Periodic checking of the picks/cm is necessary. Tiredness, variation in warp tension or simply lack of concentration can all cause the weft sett to vary and a weaver needs to be constantly aware of it.

A method of checking the sett of the cloth is to use a frame as shown in Illustration 26.

This is marked in centimetres and millimetres, and is placed on the fabric. The weaver counts the number of picks within the frame. This is easy if coarse yarns are being woven, or if the weft is striped with a regular striping pattern. It is not so easy if the threads cannot be seen individually; then a magnifying glass is needed as well. The best alternative is a linen prover or pick glass if this is available, see Illustration 27.

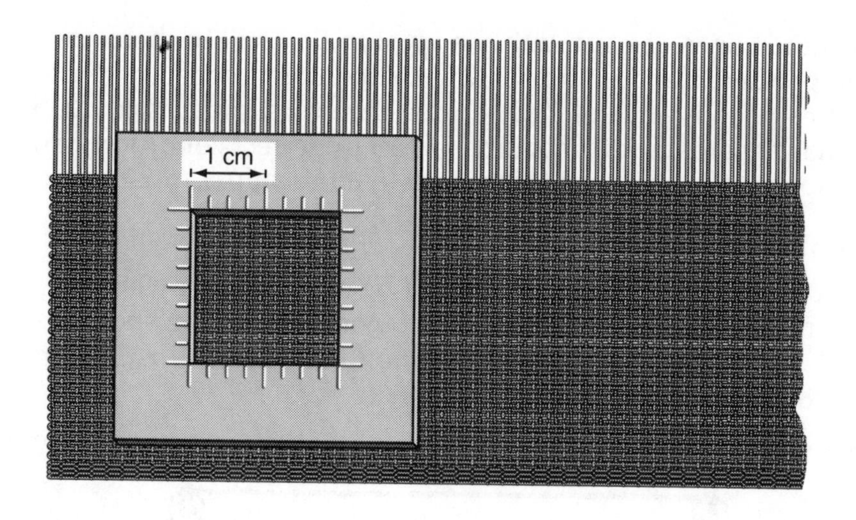

Illustration 26 Metal frame used to determine sett of cloth (actual size)

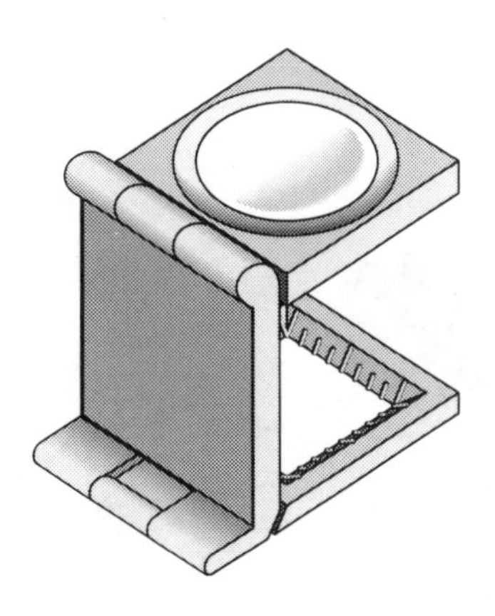

Illustration 27 Pick glass or linen prover

A straight regular selvedge is an indicator of a well-woven piece of cloth and enhances its value. Good selvedges are usually dependent on good pirn winding. A pirn is the package of weft yarn in the shuttle. The yarn must be wound under even tension in such a way that the yarn can be drawn off without snagging or catching on the pirn or shuttle. Snagging pirns causes strain and breakage of selvedge warp threads and they slow down production. Many weavers prefer to wind their own pirns. If the organization or economics of the production unit makes this impossible, it is advisable to pair up weaver and winder, that is, to ensure that a winder works as a team with a particular weaver or weavers. This makes it possible to track any snagging problems back to their source.

Knitted fabrics

Knitting machines are, on the whole, delicate machines. They require careful and sensitive handling in a clean and well-lit area. Most hand machines are not built to take the same continuous hard wear as power-operated machines.

The single most important aspect of quality control is the yarn package. This must be wound in such a way as to allow a snag-free tension-controlled yarn feed.

Machine-wound cones of yarn with the spinning oil still in them are most suitable. With a good package it should be possible to maintain a constant tension on the yarn which ensures a constant stitch size and therefore consistent quality. It is also important that needles on the machine are all in perfect condition; faulty or broken needles cannot be tolerated, and this means that the knitter must be alert at all times.

Fabric inspection

All fabric should be inspected. This should be done firstly by laying the fabric on a flat surface and searching it for faults. The second inspection should be done with a light behind the fabric. An arrangement for this second inspection is shown in Illustration 28.

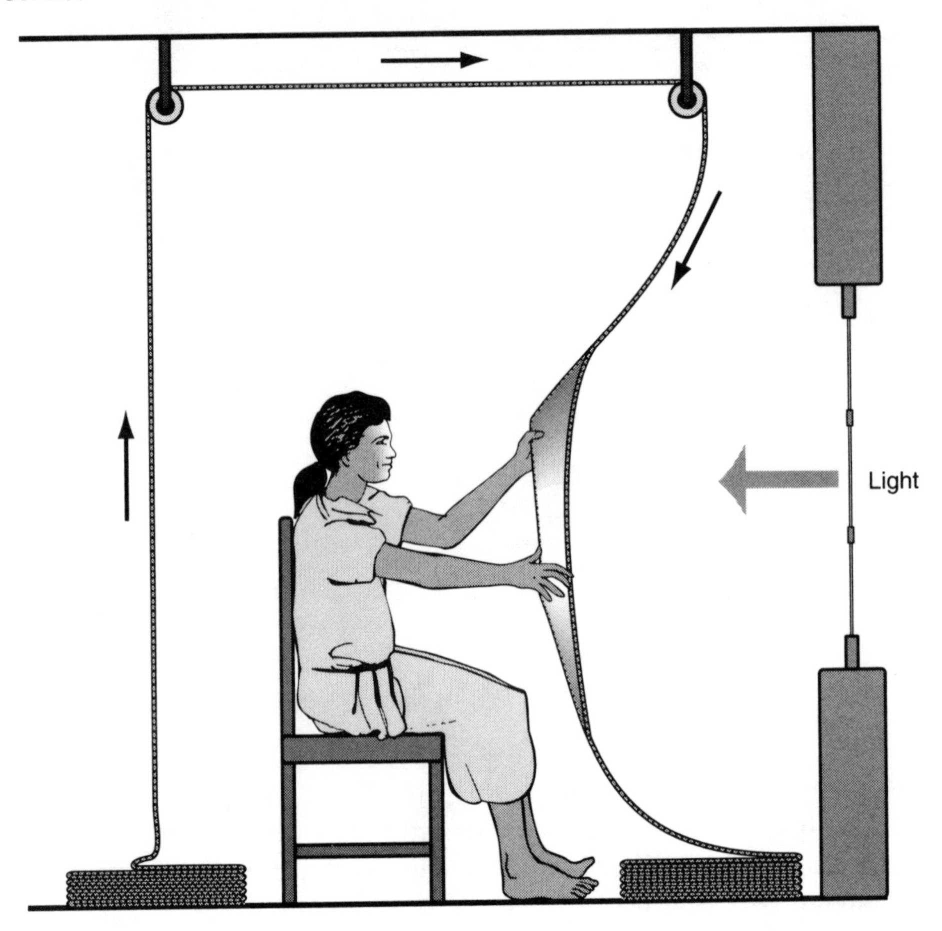

Illustration 28 Fabric inspection using natural light

The inspector sits looking at the fabric, which is gradually pulled down, showing a large area at one time with the light showing through.

Faults in woven fabrics

Weft faults. Weft bars are unwanted stripes across the fabric, differing in appearance from the remainder of the fabric. The chief cause is faulty yarn. The yarn may vary in thickness, or there may be differences in the appearance of the yarn along its length, such as colour variations caused by bad dyeing. Sometimes this is a periodic variation due to a faulty component on the spinning machine, and this results in barre, a regular pattern which is quite often diamond-shaped, see in Illustration 29. Effects such as these show up most clearly on plain fabric.

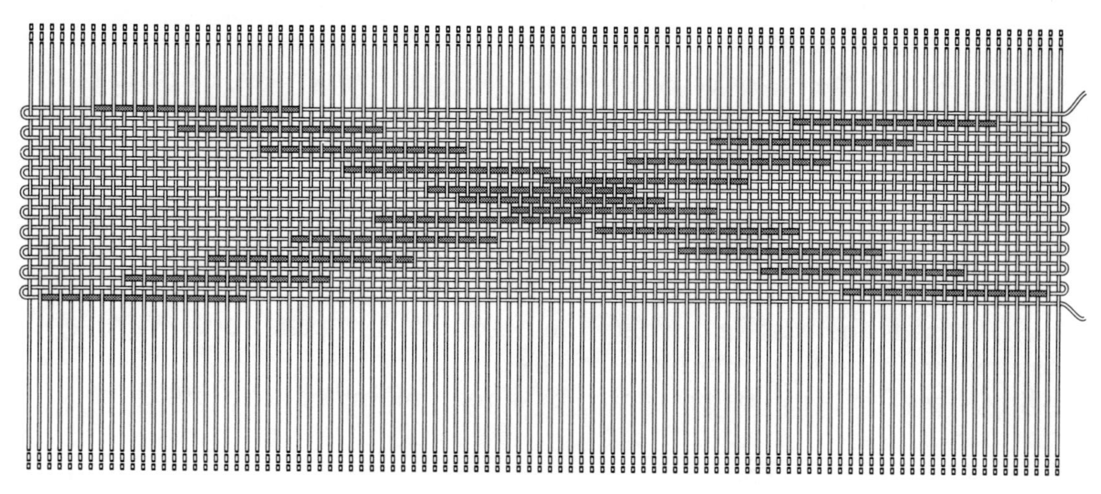

Illustration 29 Diamond barring

A single, wide, *weft band* is an area of fabric occupying the whole width, different in appearance from fabric before and after. This occurs if a pirn has been used with different yarn from that used in the remainder of the fabric. This implies faulty supply of pirns to the weaver.

The band or stripe can also be caused by a variation in weft sett. This means that the weaver has beaten up the cloth to different extents. In power-loom weaving the cause is a variation in the take-up of the fabric; on these looms a variant of this is the *let-off mark,* a defect due to a badly-adjusted let-off motion on the loom.

Faults which show as a single pick width across the loom can also occur. One is the faulty insertion of the wrong yarn, perhaps the wrong shuttle in multi-shuttle weaving, or the accidental insertion of a double pick. A *shuttle mark* is a weftwise mark which occurs when the shed is too small and the shuttle abrades the warp yarns as it passes through. Other weaving faults are the *stop mark* and *set mark,* which show as a variation in pick spacing over a short distance, occurring when the weaver has stopped weaving for some reason and then failed to get the tension right on re-commencing.

Warp faults. Faults in the warp direction are usually warp streaks caused by: faulty warp yarn; warping section marks; bands of different warp threads; gaps due to broken ends not being replaced, and slack and tight ends due to incorrect tension on the warp. This latter fault usually only appears at the beginning of the weaving as tension variations even out as the weaving continues. Another warp-way fault is the *reed mark;* gaps between ends caused by the dents in the reed separating threads. This is prevented by correct matching of the weave and denting.

Other weaving faults. Holes can occur in the fabric. There may be some damaged threads; there may be a hung pick, that is one which is caught on a knot in a warp thread that has moved forward with the knot; or there may even have been a smash, when the shuttle has collided with the warp.

Other small faults can occur due to faulty operation of the loom, for instance yarns floating instead of being interlaced correctly, when coils of yarn slip off the pirn and are woven in.

All of these faults should be recognizable, and possible recurrence prevented.

Faults in knitted fabrics

As with weaving, the major cause of faulty fabric is faulty yarn, causing barriness because of the variation in appearance or thickness of the yarn along its length. Bands can occur due to a change in the appearance of the yarn from one package to the next.

Variations in yarn tension can cause loose or tight courses, which can in turn result in courses of either long or short stitches.

Faulty needles can have several effects. One is the dropped stitch, which produces a *ladder*. There is the *miss-knit*, where the needle fails to pick up the yarn, and the *birdseye*, where the needle accidentally tucks.

All of these faults can be prevented by careful maintenance of machinery.

TESTING METHODS

Quality control involves the continuous checking of fabric while in production. Further checking can be carried out on fabric after production, in order to grade the fabric, determine the fault rate, ensure that the fabric quality meets the desired specification and that the standard of workmanship is being maintained. Even if the specification for any single fabric is not within fine limits, it is necessary to maintain that specification during the production of an order and to be able to carry out repeat orders.

It is possible to carry out a wide range of tests on fabrics to determine their properties, but if the correct yarn is being used and the fabric is made to specification, then it should have the right properties. This means that testing for quality can be restricted to tests which check that the specification is being met. Simple fabric tests include measurements of the number of threads in each area of the fabric, plus its dimensions and weight. It is essential that supplies of the correct yarn are readily available and that it is of consistent quality. This basically involves testing samples of the yarn as it is received, to ensure that it has the correct counts and is not uneven.

Suitable yarn tests have been described in the ITDG handbook, *Spinning*.

Testing woven fabrics

Determination of sett. The most useful measurement is the determination of the numbers of ends and picks per centimetre. A simple way has already been described for carrying this out on the loom; the same procedure can be used on the fabric off the loom. The number of ends and the number of picks in a unit square are counted. Preferably, the measurement should be carried out in ten places along the length of the fabric, and at a fewer number of places across the width. Do not take the measurements from within ten centimetres of the selvedge. The results should indicate whether the fabric is up to specification and if there is too much variation in the weaving.

Determination of dimensions. One straightforward measurement is the width of the fabric. The fabric should be laid out on a flat surface and a ruler or tape measure used to determine the total width of the fabric at frequent intervals along its length. The measurements should not vary by more than 1cm in a metre width.

A second measurement is the weight, which is expressed in terms of the weight per unit area.

The first method is to weigh the whole piece of fabric from the loom. The width and length of the fabric are then measured with a tape, and the weight per unit area determined by dividing the total weight by the product (fabric width x fabric length). This gives an accurate mean value for the weight per unit area, but tells you nothing about any variation along the fabric.

A more helpful measurement would be to determine the weights of samples along the fabric. One way is to take full width strips and weigh these, then dividing by the width x length of the strip. An even more useful determination is to cut ten samples, each 10cm x 10cm, and weigh these.

In both these methods the fabric has to be cut, and this may be out of the question.

Knitted fabric

Knitted fabric is made by forming yarn into loops, and consistent quality is achieved by the use of high quality yarn, while maintaining a constant loop length during knitting.

The course length is the length of yarn which is knitted into one course of the fabric. It is a good indication of quality. The length can be determined by unravelling one course from the fabric. The yarn is then held straight and its length measured. The most accurate way to get a straight yarn is to hang a weight on it. One end of the yarn is fixed on to the wall, preferably by a clip, a weight hung on the other end, and the distance along the yarn measured. The weight should be enough to straighten the yarn without stretching it. A permanent testing facility for course length can be set up by fixing a metre rule vertically on the wall with a clip for the yarn at the top, and a small weight with a clip ready to attach to the yarn.

The loop length is determined from the course length by dividing by the number of wales in the fabric or the number of needles on which the fabric was knitted.

This measurement cannot be made without unravelling the fabric, and so cannot be used during the knitting of a length. It can be used, however, to ensure that a fabric has been made to the same specification as a previously made one. It can also be used to check that the correct stitch-cam setting is being used; a small sample is knitted, taken off the machine and the course length determined.

A method of checking loop length during knitting is to mark a given length of yarn prior to knitting. A length of one metre is measured between the knitting machine and the cone feeding it, and each end of this metre is marked by pen on the yarn. The yarn is then knitted into the fabric, and the number of stitches (n) knitted from the metre is counted. This will give a value for the loop length (l), ($l = 100/n$ cm).

The reason why course length is an important guide to fabric quality is that the course length is the one aspect of fabric construction which does not change. A problem with knitted fabric is that the dimensions can change in different conditions. When removed from the machine, the loops will change shape as the fabric relaxes, and the relaxation can continue when the fabric is washed. As this happens, the numbers of courses and wales change, and the fabric may change in width and length. This can be disastrous when the fabric is made up into garments; therefore knitted fabric is usually given a wet treatment before making up. It would be useful to be able to test knitted fabric quality by measuring the number of courses per cm and wales per cm, but the results would be unreliable because they depend on conditions such as the take-down tension during knitting. The measurement of courses per cm and wales per cm can be useful when carried out under specific conditions. The fabric should be put into water and stirred to allow the fabric to relax. It should then be dried flat and measured. A previously-prepared frame, of internal measurements 10cm x 10cm, is laid on the fabric and the number of wales across and the number of courses down are measured; each divided by ten to give the values required.

4. TYPES OF HANDLOOMS AND KNITTING MACHINES

HANDLOOMS

Frame loom

The frame loom can be made almost entirely of wood although other materials can be used such as bamboo or palm tree. It is a strong and simple loom which can withstand heavy daily use over long periods. Its dimensions allow for comfortable efficient use; it is designed for the physical comfort of the weaver and the need to control the weaving process efficiently. The loom should be reasonably inexpensive to make and easy to repair. It can be constructed almost anywhere in the world.

It is best suited to the production of plain fabrics. The fabric can be produced at up to 10 metres a day at 10 picks/cm. When fitted with a fly shuttle, widths of up to 2 metres are possible; without a fly shuttle 1 metre is the comfortable limit. Warps of up to 60 metres length are possible.

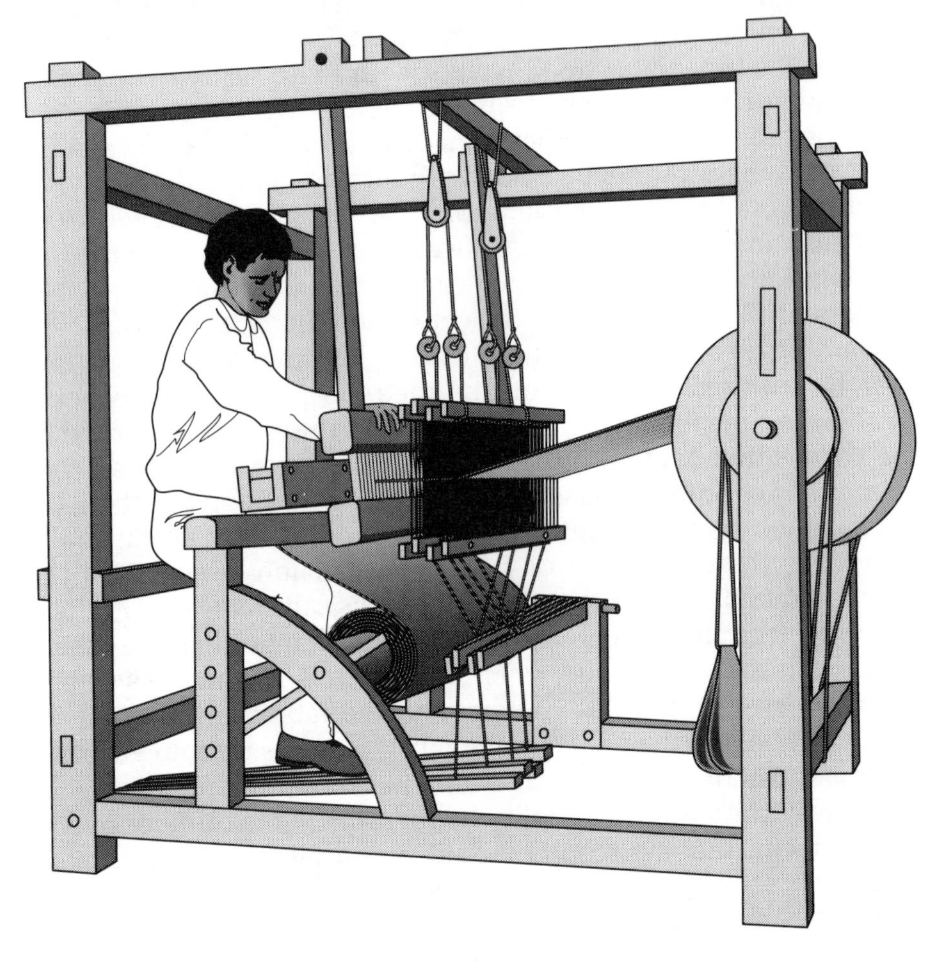

Illustration 30a Frame loom

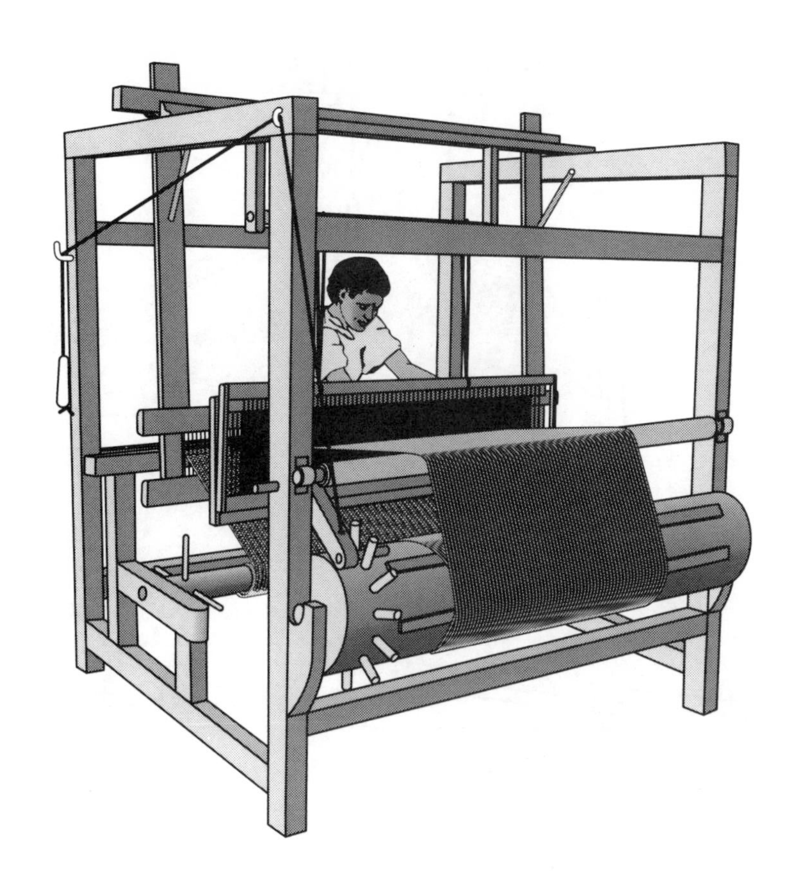

Illustration 30b   Frame loom

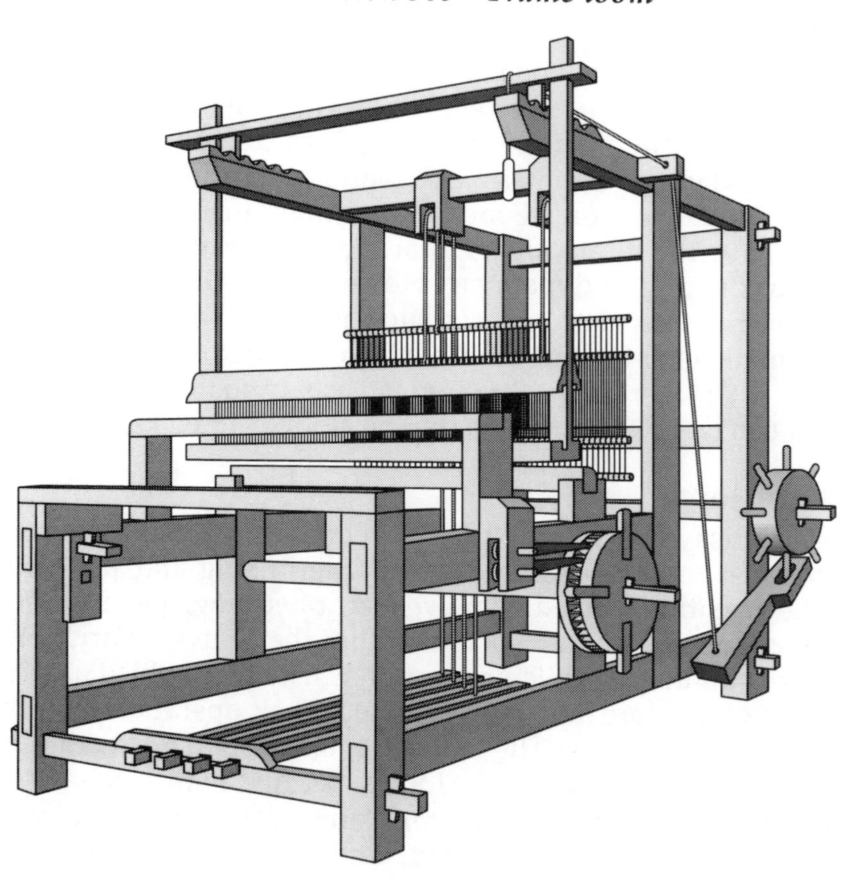

Illustration 30c   Swedish loom

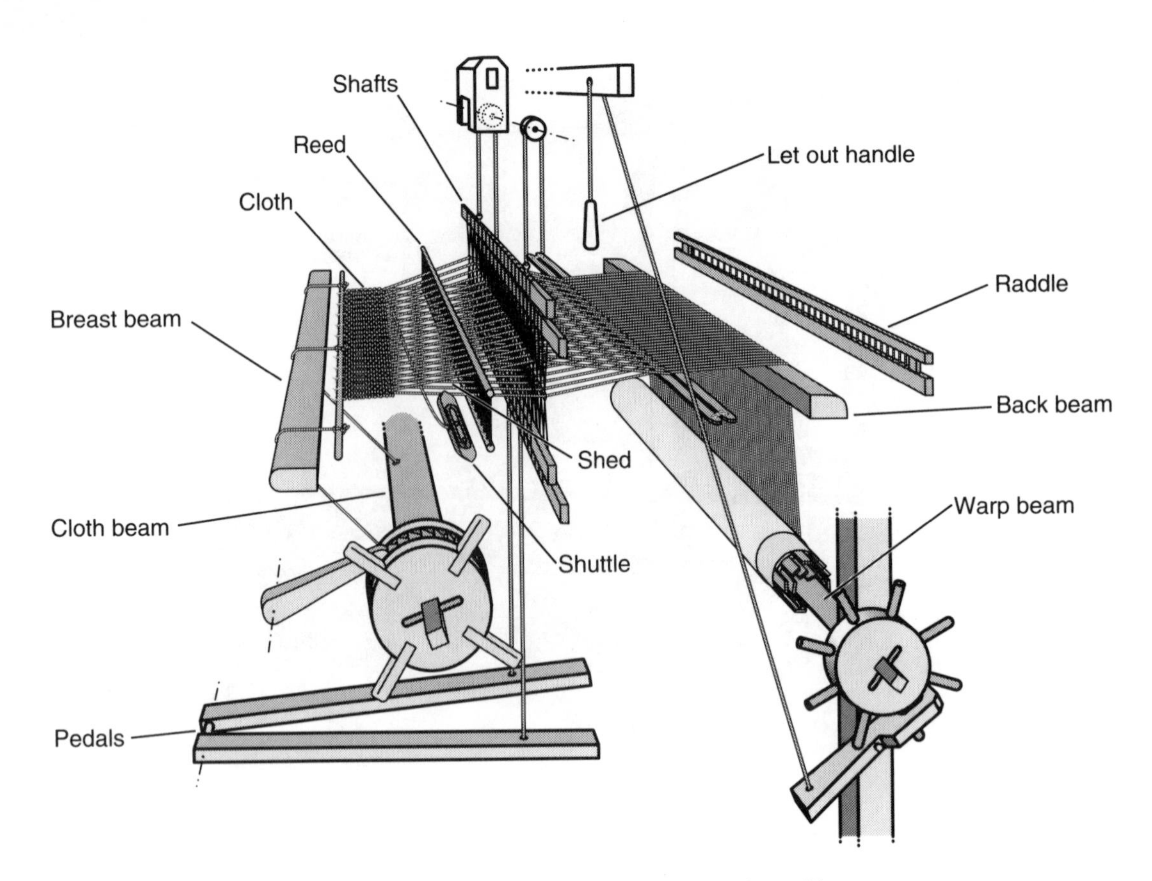

Illustration 30d   Working parts of a handloom

Pattern or dobby loom

This loom is usually made from wood and metal. It incorporates an automatic warp take-up and a multiple shuttle box batten. The shedding mechanism, called a dobby or witch, allows for large numbers of shafts (16, 24 or sometimes 36) to be handled. Unlike the counterbalance or countermarch loom, the shedding sequences are pre-selected and stored with pegs on wooden lags, simplifying the weaving procedure.

These looms have the disadvantage that they are heavy and tall, and require specialist skills to construct and maintain. However, they have been developed for the production of multiple shaft patterns such as fancy twills, brocades or double cloths. The automatic take-up and multiple boxes reduce the need for continuous stopping and starting, thereby facilitating consistent quality and increasing production. Output can be up to 6 metres a day on cloth of 16 picks/cm.

Semi-automatic loom

Many varieties of this type of loom exist. They eliminate the pick-by-pick control of the cloth by the weaver. This is achieved by interlinking and pre-setting all the functions of the traditional handloom weaver. Shedding, picking, beat-up and wind-on are all controlled by the machine rather the weaver. Variations in fabric characteristics are eliminated while production is usually doubled at least.

Semi-automatic looms are manually operated. Their operation requires less skill than the traditional handloom. The more important person in this case is the technician who pre-sets or maintains the looms. Because of the speed of the shuttle, accurate weft winding is essential. Operators can therefore expect to earn less than traditional weavers as less skill is required, while technicians will earn more. The economic balance of capital, profit and operator wages will vary in different circumstances. It will be radically different from production by more traditional methods and may limit the production applications. Most semi-automatic looms are normally made from metal although some are made of wood. Construction and maintenance are specialist jobs.

The loom is best suited to the manufacture of plain cloths or twills, and cloth can be produced at up to 20 metres a day at 16 picks/cm (Illustration 31).

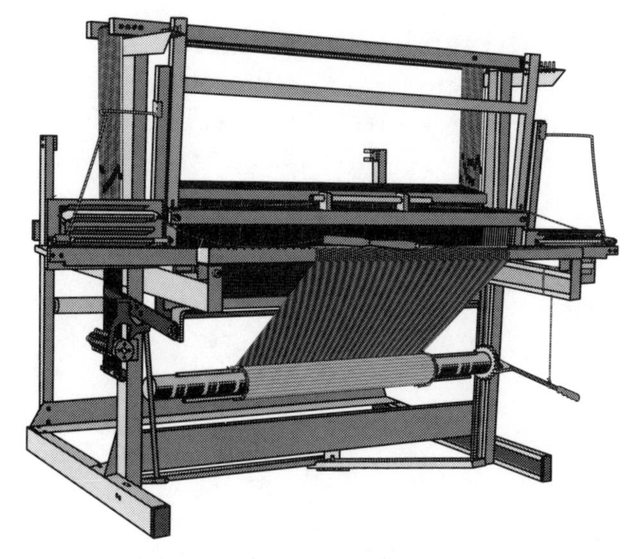

Illustration 31 Semi-automatic loom

KNITTING MACHINES

Knitted fabric can be made by hand by using two pins, but this method is highly unproductive compared with the use of a knitting machine. On the other hand, knitting machines, although fast and fairly simple to use, are complicated pieces of equipment. They normally employ latch needles, the production of which must be left to commercial manufacturers. Thus even the simplest of knitting machines have to be bought from commercial manufacturers rather than being made locally. In addition, unless coarse gauge machines are to be used, there may be a problem if there is not a ready supply of replacement needles.

Simple hand V-bed knitting machine

This is a hand-operated machine made of wood with robust latch needles which should not therefore need replacement. It is flexible in the variety of fabric structures it can produce because the knitter must position the needles by hand every course. Simple 1/1 rib fabrics can be made at a reasonable speed; the productivity goes down as the fabric structure becomes more complicated. A reasonable output of fabric would be 8 metres per day.

Hand V-bed machine

This is a hand-knitting machine with two needle beds of latch needles. It has a convertible needle bed which allows simple switching from 5.5 gauge to 2.5 gauge. Knitting is done by the movement of a carriage across the machine by hand; each movement creates one course of knitting. A wide range of patterns can be produced by resetting the cams between courses, and by transferring stitches by hand, garment shapes can be produced.

Although the machine must be obtained from a recognized dealer, its high productivity means that return on capital is fast (see Illustration 21).

Circular hand knitting machine

This is a small, hand-operated knitting machine for tube knitting. It is of metal construction with latch needles. It can be used for making socks and stockings, scarves and hats. It will only produce a single width tube in a single gauge with an output of approximately 3 metres of plain knitted tube per hour (see Illustration 20).

Small circular motor-driven knitting machine

This simple power-driven single jersey knitting machine is capable of producing narrow (2-5cm diameter) flat knitted tubes suitable for medical and sanitary protection use (Illustration 32).

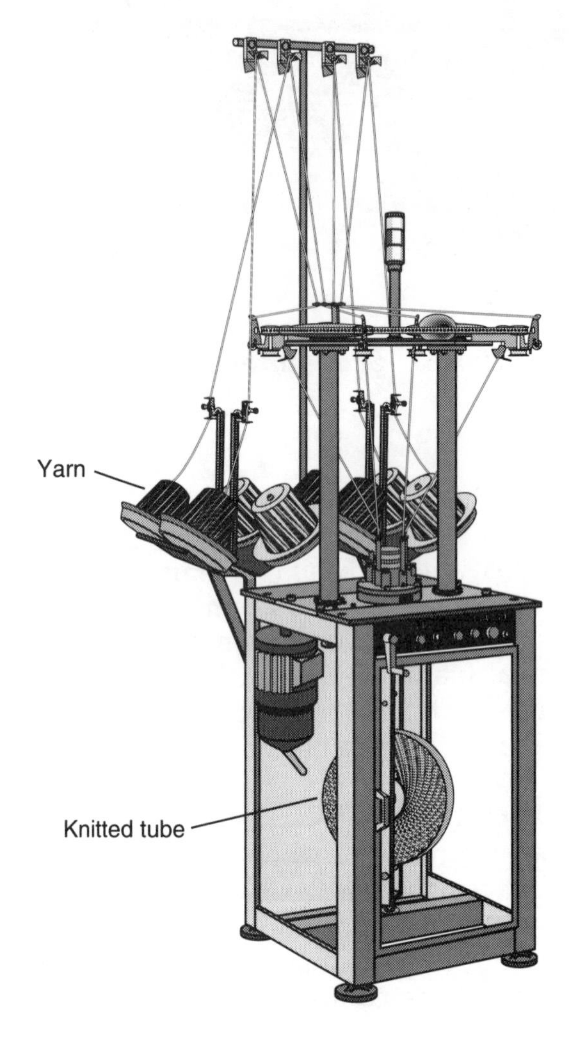

Illustration 32 Small circular, motor-driven, knitting machine

5. PLANNING FOR PRODUCTION

The circumstances in which fabrics are made on simple equipment are likely to vary widely. It may be intended to establish fabric production in a new situation, where there is little local experience in the field, or perhaps there is an established local activity with considerable expertise in a particular method of making fabric, and this is to be expanded or improved. Whatever the intention, careful planning and a few basic requirements must be met to achieve successful production.

If the fabric manufacture is to be organized in a centralized manner, rather than in individual households, the decision must be made based upon the market for the fabrics to be produced. The yarn supplies required must be available on suitable packages and to the range of counts and colours which will allow the designs to be produced. If the yarn has to be supplied from a distance, the availability and cost of transport must be carefully assessed.

An important factor is motivation. In general, it is true that machines such as looms designed for fast and efficient operation are heavier to work and require less skill to operate. These characteristics bring problems of tiredness, boredom and eventual lack of interest in the work. It is therefore important, when considering the kind of equipment to be used, to build into the organization ways of keeping up interest. Without that interest the quality of work will suffer. The use of equipment such as fly-shuttle looms or semi-automatic looms can cause problems of this nature. Self-employed workers are less bothered by this sort of problem; they can balance their effort and discomfort, mental or physical, against the profits they will make to support themselves and their families. Employers of workers, however, will find this a problem which is difficult to solve. One possibility is to permit the workforce to share in the overall financial success of the workshop or run the enterprise as a co-operative.

The way in which the work needs to be organized locally needs to be considered. The space or building in which it will take place can be arranged in any way which is most convenient but it should be dry, light and airy, with good shade. Storage of yarns and fabrics should never be in direct sunlight. The space should be arranged so that materials can be moved easily. The building should include a secure area both for yarn and fabric. The storage area should be light and have good ventilation. All materials should be stored off the ground on wooden shelves or slats, and in a way that allows frequent inspection for signs of attack by insects or mildew. All natural fibres are subject to such attacks, and in the absence of chemical treatment to prevent them, frequent inspection and turning, combined with storage in light and airy conditions out of direct sunlight are the best precautions.

A consistent feature of textile manufacture is the presence of fibre fluff. This settles on all surfaces and when it becomes mixed with lubricating oils, it can be a constant source of oily stains on the cloth. Light-coloured fabrics are particularly vulnerable. In hot climates, sweat and dust can also combine to make marks on the cloth, often on the end of yarn packages. Once on the fabric, these stains are difficult and costly to remove. Each time that equipment is empty, the opportunity should be taken to clear away fibre fluff, and rags and towels need to be made available to try to minimize the effects of oil and sweat.

Some waste always occurs during textile processing, including yarn and fabric; these should be collected and stored for resale as seconds or waste.
The area in which yarns are received should contain a bench and a suitable weighing apparatus to check on deliveries. Other areas need to be set aside for inspection and testing, and for packaging the fabric ready for delivery to the customer.

Training should be considered an essential step before production if the equipment or techniques being used are new to the area. This will be most needed for a unit with knitting machinery, but even for a weaving plant, as the looms installed may be different from those used by the workers previously.

Skill and knowledge of the process will bring dividends in terms of fabric quality. If possible, some training should be started, even if only for one or two key personnel, before the establishment of a new unit. The training programme should also include discussions with all those who might be involved with the new activity. The social problems which may arise from a failure to introduce new ideas sensitively and with local agreement, can be a major barrier to any new venture. Suitable training is sometimes available at major textile centres or educational institutions in the area and these possibilities should be investigated. Any training programme must also include the health and safety aspects of using new and perhaps unfamiliar equipment.

Before making any decisions about the scale of equipment needed in a new situation, the following points should be carefully considered.

Costs

Is there sufficient justification or need for the level of expenditure planned? Is there sufficient cash available to meet purchase costs? This includes not only the cost of the basic equipment, but a stock of spare parts, ancillary equipment, wages, day-to-day running expenses and yarn stocks to make fabric prior to the initial sales. If money is borrowed for purchases, can it be repaid satisfactorily? Is there government or non-governmental funding available to cover the capital cost of equipment or for setting up small industries?

Capacity

Does the choice of equipment match the desired fabric production level and the available yarn supplies? Is there room for expansion? Is the equipment flexible enough to give some variation in fabric design and fabric weight?

Location

If all the equipment is located in one place, will this suit all those who are expected to use it? Would temporary locations and portable equipment be more suitable? If materials have to be moved between different locations, is suitable transport available at reasonable cost?

Type

Is the type of equipment suitable for the fabric which has been chosen as marketable? Will it make a range of fabrics allowing for changes in the market or commissions for different fabrics? Will the equipment stand up to daily use, year in, year out? Heavy wooden frames, held together with mortice, tenon and pegs, easily repaired and adapted are preferable to light ones unless portability is required.

Availability

Is the equipment available locally from a reputable manufacturer? If not, is anyone willing to manufacture or sell the equipment in the district? Are manufacturing instructions available and are there sufficient skills to carry out the construction? What kind of spare parts and advisory service is on hand or will be needed? If power-driven equipment is being planned, is a reliable power supply available for a reasonable period each day?

Experience

Is the equipment easy to use? Is the technology easy to develop new fabric designs for? Can tuition be obtained locally or at a national training centre? What local skills exist, or could be learned, for maintaining and servicing the equipment?

Social acceptability

Will the choice of equipment and system be readily acceptable to local people? What changes to existing social practice will be required? Has the demand for change come from the local community or from outside? After shortlisting the most acceptable range of equipment, an economic evaluation of all inputs and outgoings associated with it will help to identify the best final choice. Finally, seek expert advice about the equipment and how you propose to use it, before purchase.

Design

Textile fabrics depend for a ready sale on the quality of their design. This means that there should be a market for the type of fabric that is intended to be manufactured, and there should ideally be a member of the staff or organizer who has the knowledge to ensure that the market is maintained, by keeping the quality of design high and appealing to the potential customer. It may be that there is a traditional local fabric and people who are well-acquainted with the production of the designs within this tradition. Even if this is so, there is likely to come a time when the range of fabrics needs to be expanded, or changes to the design are required to fulfil customer requirements. This all means that there should be the potential to change fabric designs at some point.

It is sometimes possible for weavers or knitters to incorporate their own designs into the fabric they are producing. It is advisable to have trained staff available who are experienced in creating new designs, have a sense of colour and appreciation of the markets. They should be aware of the colours fashionable at any one time, and be able to translate their designs into instructions for the weavers or knitters involved in fabric production. This may mean the employment of several people with different skills, or of one person who is capable of running all the design operations.

New designs might also be submitted to the weavers or knitters, who should be capable of translating them into fabric, by a customer who is in closer contact with the market place.

Economics of fabric production

Once the operation has been planned in some detail, it is a useful exercise to undertake a trial costing for the production of fabric to give a rough idea of the viability of the operation in commercial terms. The cost of a fabric has two components: overheads (fixed or indirect costs), and variable (direct costs).

Indirect costs

- ☐ Interest on the cost of any stock of raw materials.
- ☐ Cost of premises.
- ☐ Heat, lighting, power and other services.
- ☐ Telephone.
- ☐ Cost of depreciation of equipment and interest on any loans for purchase.
- ☐ Consumable materials.
- ☐ Insurance.
- ☐ Stationery and postage.

Direct costs

- ☐ Raw materials (yarns, dyes, chemicals and labels).
- ☐ Wastage.
- ☐ Transport.
- ☐ Wages (including any contribution to a welfare fund and any incentive wages).

Once an estimate of production has been made for the planned unit, the above costs can then be worked out for, say, a week's production. The actual fabric costs are then the weekly indirect costs plus the weekly direct costs, divided by the length of fabric produced each week. To this actual cost must be added any profit per metre of fabric to give the final selling price.

Productivity

One of the important features of the weaving loom is the fact that it is capable of producing a range of different fabrics.

Fabrics such as pile fabrics like velvet, and heavy industrial fabrics like conveyor belts may require special loom modifications, but in general there is a considerable choice open to the weaver.

The knitter has less scope for variation in construction. In general, for knitting and weaving, the more complicated the structure being produced, the lower the speed of production. However, this is balanced by the fact that the price obtainable for a more sophisticated product may outweigh the disadvantage of lower productivity.

Influence of construction on production rate

The rate at which fabric is produced depends on the construction. For instance, in weaving, it depends upon the weft sett (the number of picks per cm). This means that if the weaver inserts picks at a rate of 30 picks per minute, the production rate will be 1cm per minute for a fabric of 30 picks per cm, while for fabric of only 15 picks per cm, the production rate will increase to 2cm per minute. Appendix 1 shows a list of woven cotton fabric specifications.

As a general principle, and as the table in Appendix 1 shows, light fabrics will have finer yarns and higher warp and weft setts than heavier fabrics. This means that they will be more difficult to produce, with more likelihood of yarn breakage, a lower production rate and a higher raw material cost.

The following table shows the length of fabric that might be expected to be produced per day by one operator working constantly. In practice, breakdowns will occur and even if not, the operator will require breaks from production, which affect the rate.

Equipment	*Fabric*	*Length (m)*
Frame loom	16 picks/cm	5
Pattern loom	16	6
Semi-automatic loom	16	20
Simple hand knitting machine	5 courses/cm	5
Domestic knitting machine	5	40
Machine powered domestic knitting	5	80

Illustration 33 shows a plan for a small weaving unit of four to ten handlooms, with yarn and fabric storage and a small area for testing.

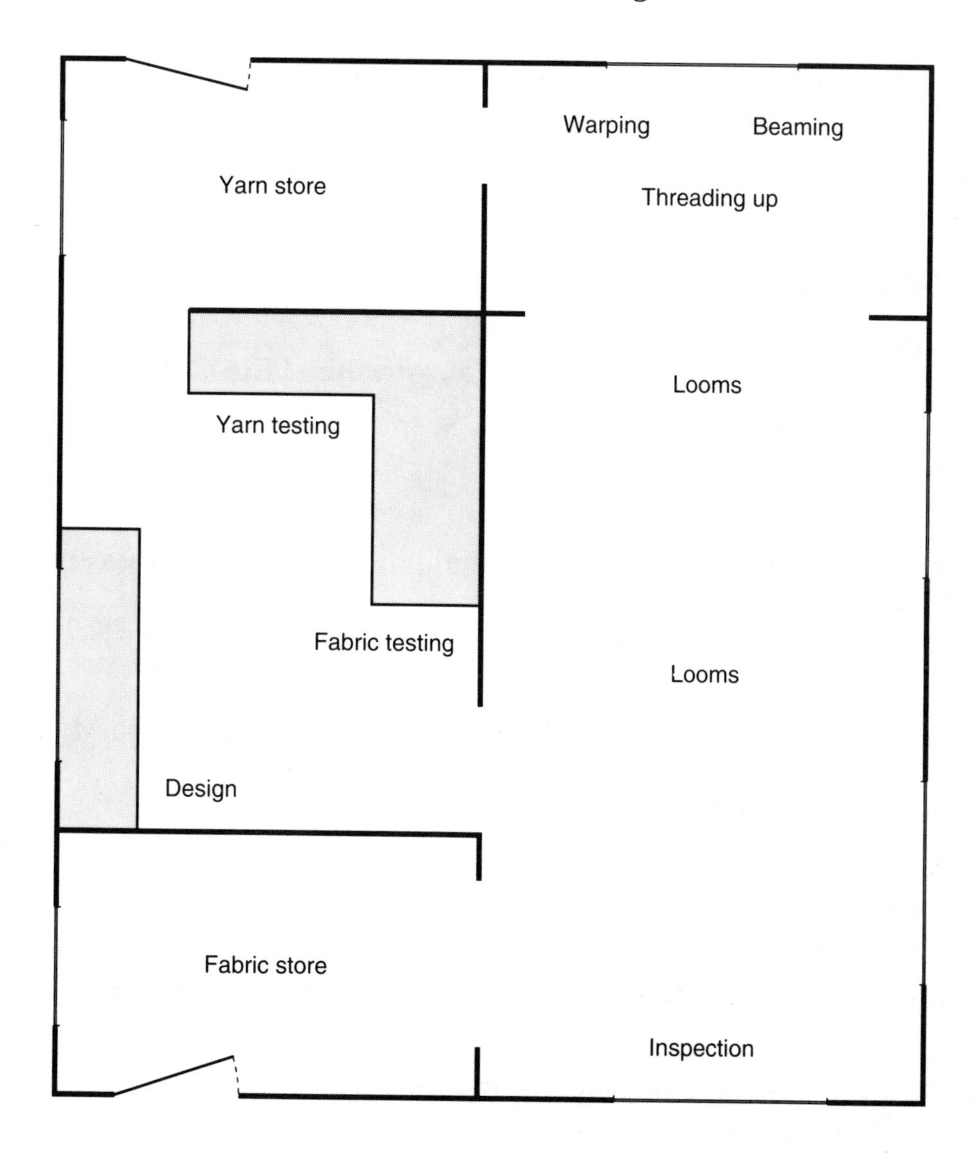

Illustration 33 Plan of a small handloom weaving unit

6. EQUIPMENT SUPPLIERS

HAND LOOMS

United Kingdom

AVL Looms Ltd, St George's Mill, St George's St, Macclesfield Cheshire
(also in USA: 601 Orange St., Chico, California 95926)

Emmerich (Berlon) Ltd, Wotton Road Ashford, Kent TN23 2JY

Dryad, Northgates, Leicester

Bonas Griffith Ltd, 12a Southwick Industrial Estate, Sunderland SR5 3TX

Sweden

Berga Hemslojdens Ullspinneri AB, S-783 02 Stora, Skedvi

India

V.P.F. Testing Equipment, Ventakapathy Foundry, Peelameedu,
Coimbatore 641004, Tamil Nadu

M/S Kamal Metal Industries, Gajjar House, Astodia Road,
Ahmedabad 380001, Gujurat

KNITTING MACHINES

United Kingdom

Frame Knitting Ltd, PO BOX 21, Oakham, Leicestershire LE15 6XB

Jones & Brother, Knitting Machine Division, Shepley Street,
Audenshaw, Manchester M34 5JD

Switzerland

Edouard Dubied & Cie S.A. (Dubied hand and power knitting machines)
CH - 2001 Neuchâtel
MADAG Maschinen - und Apparatebau Dietikon AG, (Passap) CH-8953 Dietikon

Italy

Semexco di H. Gierlinger & C s.n.c., 20090 Trezzano s/n (Milano), Via Morona, 79

TESTING EQUIPMENT

United Kingdom

Shirley Developments Ltd, PO Box 162, Crown Royal, Shawcross St.,
Stockport SK1 3JW

WIRA Technology Group Ltd, Wira House, West Park Ring Road, Leeds LS16 6QL

India

V.P.F. Testing Equipment, Ventakapathy Foundry, Peelameedu,
Coimbatore 641004, Tamil Nadu.

M/S Kamal Metal Industries, Gajjar House, Astodia Road,
Ahmedabad 380001, Gujurat.

7. SOURCES OF FURTHER INFORMATION

RESEARCH ORGANIZATIONS

Research organizations almost always have an information service on a wide range of textile topics. This usually includes book-lists and copies of research papers.

They will usually undertake testing or other work, for which a charge is made.

They sometimes organize courses or training programmes in aspects of textile manufacture.

United Kingdom

International Wool Secretariat, Technical Centre, Valley Drive, Ilkley, LS29 8PB. Tel. 0943601555. Telex 51457

British Textile Technology Group (BTTG):

> British Textile Technology Group (BTTG), Shirley Towers, Didsbury, Manchester M20 8RX.
> Tel: 061 445 8141. Tlx: 668417. Fax: 061 434 9957.
>
> British Textile Technology Group (BTTG), WIRA House, West Park Ring Road, Leeds LS16 6QL.
> Tel: 0532 781381. Tlx: 557189. Fax: 0532 304195.
>
> British Textile Development Group (BTTG), Newton Business Park, Talbot Road, Hyde, SK14 4UQ.
> Tel 061 367 9030. Fax 061 367 8845.

India

The South India Textile Research Association (SITRA), Coimbatore 641014, Tamil Nadu

Ahmedabad Textile Industry Research Association (ATIRA), Polytechnic Post Office, Ahmedabad 380015, Gujarat

Textile and Allied Research Association (TAIRO), Baroda, Gujurat
For other contact addresses, see the ITDG handbook, *Spinning*.

PUBLISHED SOURCES OF INFORMATION

Books about textiles containing information on fabric manufacture, some of which can only be obtained from libraries (marked with an asterisk *).

Hall, A.J., *Standard Handbook of Textiles* (Butterworth & Co.)

* Shaw, C., and Eckersly, F., *Cotton* (Pitman & Sons)

Brearley, A., *The Woollen Industry* (Pitman & Sons)

Atkinson, R.R., *Jute* (Temple Press)

Von Bergen, W.(Ed), *Wool Handbook* (Interscience)

Weaving

Brown, R., *The Weaving, Spinning and Dyeing Book* (Routledge and Kegan Paul)

Emery, I., *The Primary Structure of Fabrics* (Washington D.C.)

Albers, A., *On Weaving* (Wesleyan University Press, Middleton, Conn.)

Atwater M. M., *Byways in Handweaving* (New York)

Black M., *New Key to Weaving* (New York)

Collingwood P., *Textile and Weaving Structures* (B.T.Batsford Ltd)

Benson A. and Warburton N., *Looms and Weaving* (Shire Publications Ltd)

Small-scale Weaving, Technology Series, Technical Memorandum No.4 (I.L.O)

Murray R., *The Essential Handbook of Weaving* (Bell and Hyman)

Weaving and Associated Processes (The Lodz Textile Seminars No.4 (UNIDO)

Marks R. and Robinson A.T.C., *Principles of Weaving* (Textile Institute)

Regenstein, E., *The Art of Weaving* (Van Nostrand Reinhold)

Grasett, K., *Setting up a Loom* Vol. 3 (London School of Weaving)

Hindsom, A.M.C., *Designer's Drawloom: An introduction to Drawloom Weaving and Repeat Pattern Planning* (Faver)

Broudy, E., *The Book of Looms* (Studio Vista)

Lamb V. and A., *Au Cameroun-Weaving-Tissage* (Ronford Books) (text in French and English)

Knitting

Spencer D.J., *Knitting Technology* (Pergamon Press)

Dubied Edouard & Cie S.A., *Knitting Manual* (Neuchatel, Switzerland)

Cooke W.D., *Knitting Technology and the Production of Medical Textiles in Developing Countries, Medical Textiles for Developing Countries,* (ITDG)

Wignall H., *Knitting* (Pitman)

Thomas, M., *Mary Thomas's Knitting Book* 3rd ed. (Hodder & Stoughton)

Smirfitt, J.A., *Introduction to Weft Knitting* (Merrow Publications)

Knitting, Lodz Textile Seminars No.3 (UNIDO)

Weaver, M., *Machine Knitted Skirts* (Weaverknits Ltd Publications)

Textile testing

* *Methods of Testing for Textiles* (British Standards Institution)

BSI Standards: particularly 1930, 1931, 2576, 2861-6, 6395, 5441(Knitting), (British Standards Institution, BSI Enquiry Section, Linford Wood, Milton Keynes, MK14 6LE, UK)

1989 Annual Book of ASTM Standards, Volumes 0701 and 0702, Textiles - Yarns, Fabrics and General, (American Society for Testing and Materials)

Booth, J. E., *Principles of Textile Testing* (National Trade Press)

APPENDIX 1
TYPICAL WOVEN FABRIC CONSTRUCTIONS

Fabric	Ends		Picks		Yarn Counts					Weave	Weight
	/cm	/in	cm/	in/	Warp tex	Warp cc	Weft tex	Weft cc		g/m²	oz/yd²
Cotton											
Voile	24	60	23	58	11.8	2/100	11.8	2/100	Plain	27	0.8
Muslin	27	68	27	68	18	33	18	33	Plain	68	2
Print Cloth	28	72	24	62	16	36	14	42	Plain	89	3
Poplin	48	122	28	72	15	2/80	15	40	Plain	110	3.3
Cambric	18	46	18	46	45	13	45	13	Plain	170	5.3
Flannelette	36	92	18	46	27	22	42	14	Plain	178	5.3
Tablecloth	29	73	24	60	26	22	37	16	Plain	180	5.4
Sheeting	24	60	24	60	35	17	33	18	Plain	183	5.5
Tent Duck	20	50	20	50	45	13	49	12	Plain	203	6
Canvas	32	80	16	40	37	2/32	100	2/12	2/1twill	306	9.2
Gabardine	45	114	25	64	27	22	33	18	2/1twill	195	5.8
Jean	37	94	21	54	37	16	37	16	2/1twill	246	7.4
Drill	34	86	20	50	49	12	49	12	3/1twill	277	8.3
Denim	32	82	18	47	45	13	54	11	3/1twill	314	9.3
Wool					**tex**	**wc**	**tex**	**wc**			
Summer Suit	22	56	20	51	44	2/44	40	2/44	Plain	220	6.5
Wool herringbone	24	61	23	58	49	2/36	49	2/36	2/2twill	253	7.6
Wool gabardine	44	110	22	56	32	2/48	32	2/48	2/2twill	260	7.8
Dress	39	98	22	56	49	2/36	49	2/36	2/2twill	327	9.8
Suiting	26	66	24	60	59	2/30	59	2/30	2/2twill	345	10.2
Serge	30	76	24	60	52	2/34	52	2/34	2/2twill	360	10.6

APPENDIX 2
COMPARISON OF TEX WITH OTHER COUNT SYSTEMS

English cotton count and tex

COTTON	TEX	COTTON	TEX
1	590	32	18.4
2	296	34	17.4
3	196	36	16.4
4	148	38	15.6
5	118	40	14.8
6	98.4	44	13.4
8	73.8	46	12.8
9	65.6	48	12.4
10	59	50	11.8
11	53.6	52	11.4
12	49.2	56	10.6
14	42.2	60	9.8
16	37	64	9.2
18	32.8	68	8.6
20	29.6	72	8.2
22	26.8	76	7.7
24	24.6	80	7.4
26	22.8	86	6.8
28	21	92	6.4
30	19.6	100	5.9

Woollen count (Yorkshire skeins woollen) and tex

WOOLLEN	TEX	WOOLLEN	TEX
6	320	27	72
7	280	28	70
8	240	29	66
10	195	31	62
11	175	32	60
12	160	33	58
13	150	34	56
14	140	35	56
15	130	36	54
16	120	37	52
17	115	38	51
18	107	39	50
19	102	40	48
20	96	42	46
21	92	44	44
22	88	46	42
23	84	48	40
24	80	50	39
25	78	52	37
26	74	56	35

Metric count and tex

METRIC	TEX	METRIC	TEX
2	500	60	16.7
4	250	64	15.6
6	167	68	14.7
8	125	72	13.9
10	100	76	13.1
12	83.3	80	12.5
14	71.4	90	11.1
16	62.5	100	10
18	55.6	110	9.1
20	50	120	8.3
24	41.7	130	7.7
28	35.7	140	7.1
32	31.3	150	6.7
36	27.8	160	6.3
40	25	170	5.9
44	22.7	180	5.6
48	20.8	190	5.3
52	19.2	200	5
56	17.9	210	4.8

INDIRECT FIXED-WEIGHT SYSTEM

English cotton count = Number of 840-yard hanks per pound

Galashiels woollen count = Number of 300-yard hanks (cuts) per 24oz

Yorkshire skeins woollen (Y.S.W.) count =
Number of 256-yard hanks per pound

Worsted count = Number of 560-yard hanks per pound

Linen count = Number of 1,000-metre hanks per kilogram

DIRECT FIXED-LENGTH SYSTEM

Tex count = Number of grams per 1,000 metres

Jute count = Number of pounds per 14,400 yards

Denier count = Number of grams per 9,000 metres

APPENDIX 3
CLOTH (WORSTED) SETTING CHART: YARN COUNT/CM

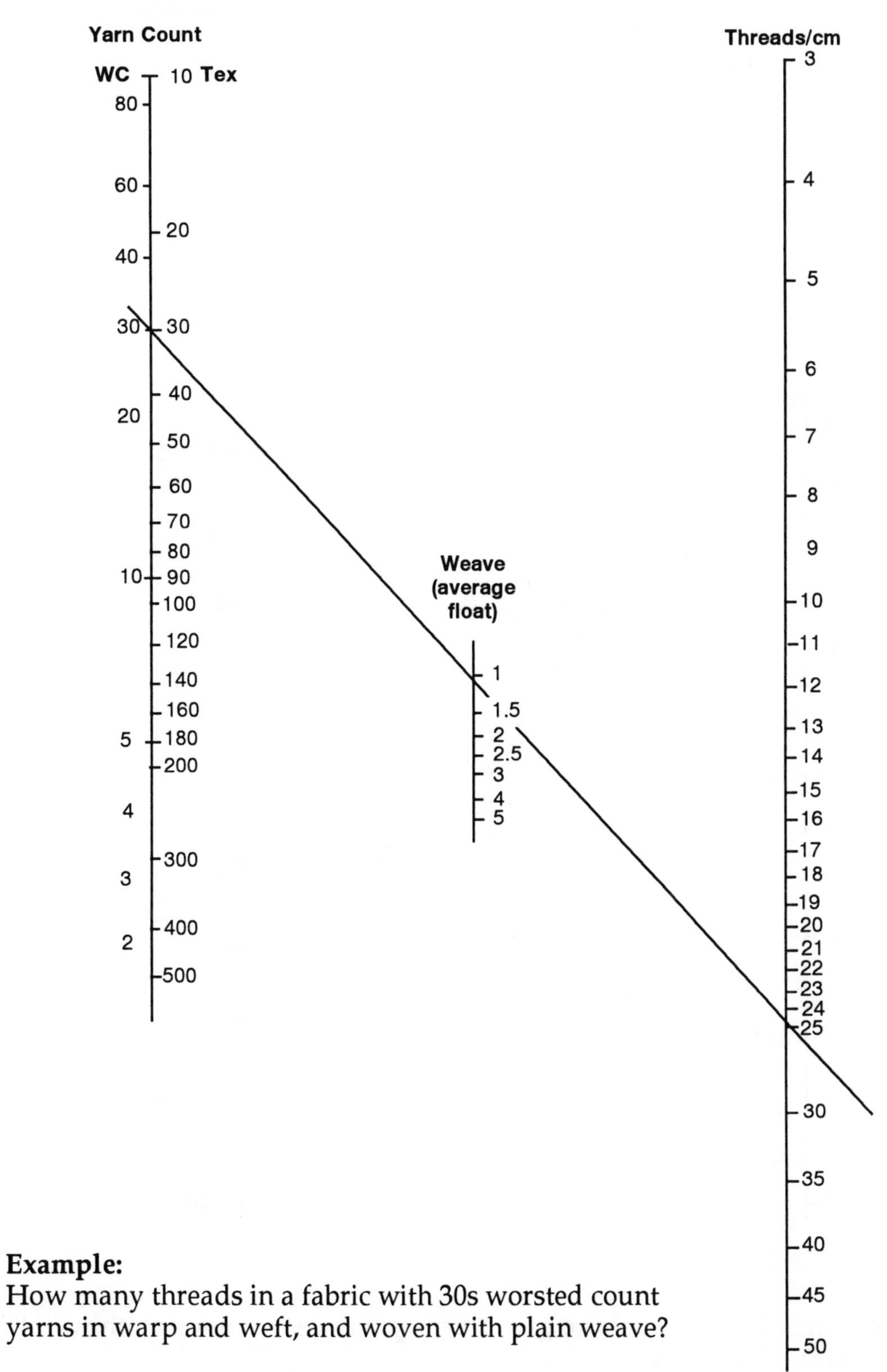

Example:
How many threads in a fabric with 30s worsted count yarns in warp and weft, and woven with plain weave?

Answer: 25 ends and picks per centimetre.

APPENDIX 4
CLOTH (COTTON) SETTING CHART: YARN COUNT/IN

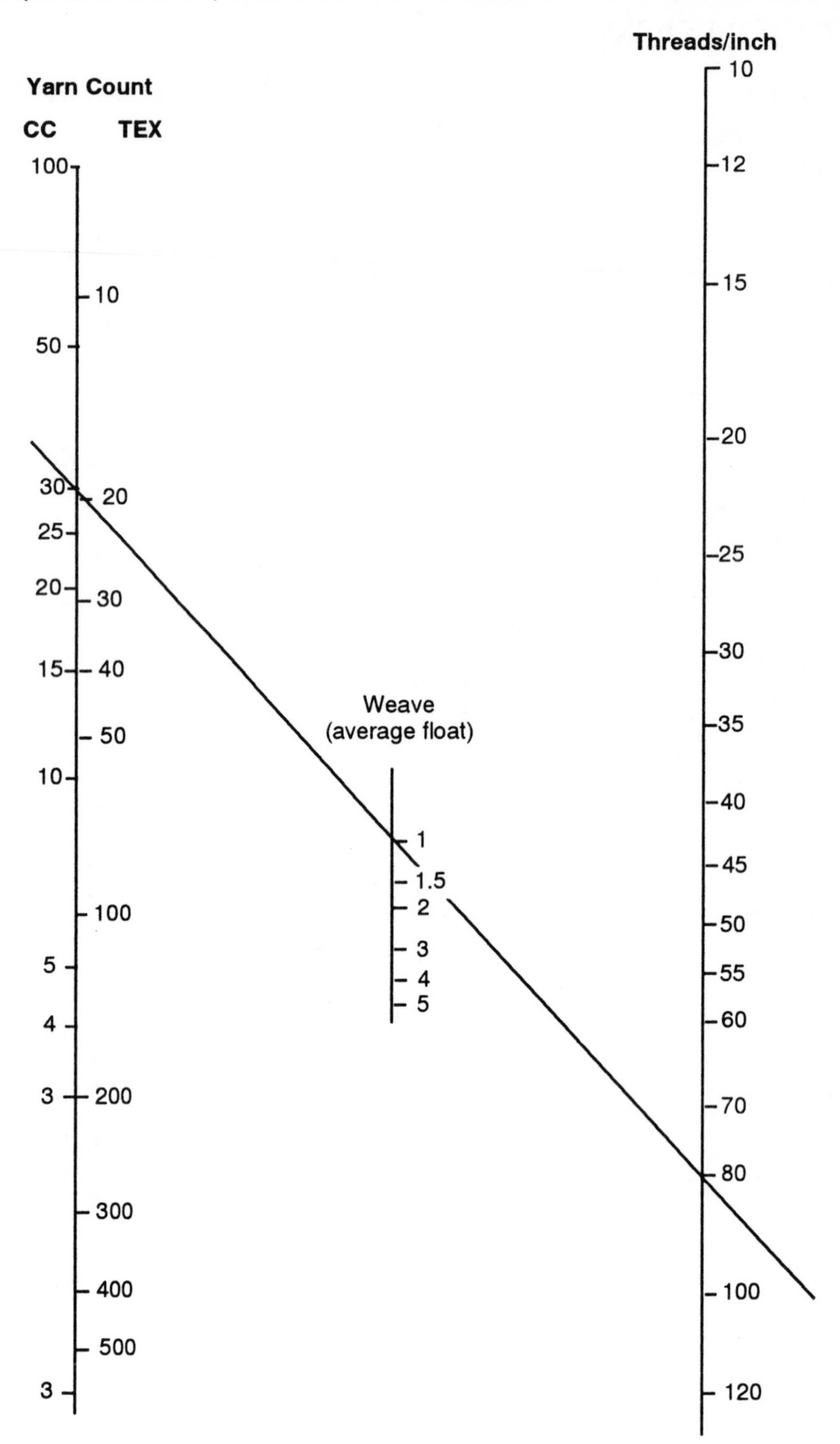

Example:
What yarn counts should be used to give a firm, plain weave cotton cloth with 80 ends/inch and 80 picks/inch?

Answer: 30s cotton count in warp and weft.

APPENDIX 5
GLOSSARY

Bar (barry or barriness)
A mark in the form of a bar across the full width of a piece of woven cloth which differs in appearance from the rest of the cloth. Often a mistake in weaving caused by either incorrect picking, wrong weft yarn or tension.

Batten
The frame containing the *reed* which the weaver swings towards and away from himself when beating up the weft into the *fell* of the cloth. Known also as a *sley* or *beater*.

Beam
A cylinder of wood or metal with bearings at each end for mounting into suitable flanges, one beam at the front and one beam at the rear of the loom. A double beam refers to two beams which can be fixed to the rear of the loom when two warps are taken up in the weaving under two different tensions.

Beer
A group of forty warp threads.

Cover
The evenness of warp spacing, giving a uniform effect on the surface of the cloth. Also the degree to which the underlying structure of a fabric is concealed by finishing.

Counting glass
A magnifying glass mounted in a small hinged metal frame with a fixed focus, the base having an aperture measuring either one square inch or one square centimetre. Used for counting the *ends* and *picks*, courses and wales in a fabric. Also known as a linen prover or piece glass. See page 27.

Dent
The spaces between the reed wires. See *reed*.

Doup
Half-heddle used in leno or gauze weaving.

End
A single warp thread through the length of the cloth.

Fell
The edge of the cloth facing the *reed* during weaving, where the last pick has been put across the warp and beaten up.

Handle
The 'feel' of a cloth.

Heald (or heddle)
A cotton cord or twisted wire, with loops at both ends, or flat stainless steel strip, each with an eye in the centre through which a warp end is threaded. Several healds or heddles are supported on a frame, harness or shaft. The frames are manipulated up and down during weaving to form the structure of the cloth.

Lags
Wooden slats into which metal or wooden pegs are secured and linked together to form a chain. The chain of lags provides the pattern information to the dobby or witch mechanism which controls the rising of the shafts to form the cloth.

Latch needle
A knitting machine needle at the top of which is a hook closed by a pivoting latch.

Package
This refers to the yarn package, being a cylindrical object made of wood, metal, paper or plastic onto which yarn is wound. Bobbins, cones, cops, dobbins, cheeses and spools are all yarn packages (See IT handbook, *Yarn preparation* by John Iredale).

Pick
A single weft thread across the width of the warp.

Pirn
A pirn is a cylindrical object made of wood, paper or plastic onto which weft yarn is wound and then secured in the shuttle for weaving. See *package*.

Raschel
A form of warp knitting, producing a fabric resembling lace.

Reed
Evenly spaced wires, steel or bamboo strips held top and bottom between baulks. Used to divide the warp threads in the loom. The reed fits into the *batten, sley* or beater. See illustration on page 11.

Reed marks
These are lines running through the length of the cloth caused by reed wires when ends are mis-dented.

Selvedges
The edges of a woven fabric.

Sett
The number of ends and picks in a square unit of cloth.

Shaft
Or harness. A parallel pair of wooden sticks between which are suspended cotton heddles or a wood or metal frame supporting the healds, through which the warp ends are threaded in order.

Shuttle box
A box at one end of the *sley* from which the shuttle is propelled to a box at the opposite end of the *sley* during weaving. See illustration on page 16.

Sley
Also known as a batten or beater. The frame supporting the reed through which the warp ends are threaded in order and is pushed backwards and forwards by the weaver to beat up the picks into the fell of the cloth. See illustration 10.

Smash
A breakage in the warp during weaving.